# *Howard Aiken: Portrait of a Computer Pioneer*

**History of Computing**

I. Bernard Cohen and William Aspray, editors

editorial board: Bernard Galler, J. A. N. Lee, Arthur Norberg, Brian Randell, Henry Tropp, Michael Williams, Heinz Zemanek

William Aspray, *John von Neumann and the Origins of Modern Computing*

Charles J. Bashe, Lyle R. Johnson, John H. Palmer, and Emerson W. Pugh, *IBM's Early Computers*

Paul E. Ceruzzi, *A History of Modern Computing*

I. Bernard Cohen, *Howard Aiken: Portrait of a Computer Pioneer*

I. Bernard Cohen and Gregory W. Welch, editors, *Makin' Numbers: Howard Aiken and the Computer*

John Hendry, *Innovating for Failure: Government Policy and the Early British Computer Industry*

Michael Lindgren, *Glory and Failure: The Difference Engines of Johann Müller, Charles Babbage and Georg and Edvard Scheutz*

David E. Lundstrom, *A Few Good Men from Univac*

R. Moreau, *The Computer Comes of Age: The People, the Hardware, and the Software*

Emerson W. Pugh, *Building IBM: Shaping an Industry and Its Technology*

Emerson W. Pugh, *Memories That Shaped an Industry*

Emerson W. Pugh, Lyle R. Johnson, and John H. Palmer, *IBM's 360 and Early 370 Systems*

Dorothy Stein, *Ada: A Life and a Legacy*

Maurice V. Wilkes, *Memoirs of a Computer Pioneer*

# *Howard Aiken: Portrait of a Computer Pioneer*

I. Bernard Cohen

The MIT Press
Cambridge, Massachusetts
London, England

© 1999 Massachusetts Institute of Technology

All rights reserved. No part of this book may be reproduced in any form by any electronic or mechanical means (including photocopying, recording, or information storage and retrieval) without permission in writing from the publisher.

Set in New Baskerville by Wellington Graphics.

Printed and bound in the United States of America.

Library of Congress Cataloging-in-Publication Data

Cohen, I. Bernard, 1914–
    Howard Aiken : portrait of a computer pioneer / I. Bernard Cohen.
      p. cm.—(History of computing)
    Includes bibliographical references and index.
    ISBN 0-262-03262-7 (hc : alk. paper)
    1. Aiken, Howard H. (Howard Hathaway), 1900—1973. 2. Computer engineers—United States—Biography. 3. Computers—History.
I. Title. II. Series.
QA76.2.A35C65 1999
004′.092—dc21
[b]                                                                          98-43965
                                                                                 CIP

for Peter Galison and Jed Buchwald, each setting standards for the history of Science, and for Maurice V. Wilkes, who has taught me so much about computers and computer history

# Contents

Preface    xi
Acknowledgments    xv
The Names "ASCC" and "Mark I"    xix

## 1
**Introduction to a Pioneer**    1

## 2
**Early Life and Education**    9

## 3
**A Harvard Graduate Student**    21

## 4
**First Steps Toward a New Type of Calculating Machine**    33

## 5
**An Unsuccessful Attempt to Get the Machine Built**    39

## 6
**Seeking Support from IBM**    45

## 7
**The Proposal for an Automatic Calculating Machine**    53

## 8
*Aiken's Background in Computing and Knowledge of Babbage's Machines*   61

## 9
*Planning and Beginning the Construction of the Machine*   73

## 10
*How to Perform Multiplication and Division by Machine*   87

## 11
*Construction of the Machine*   95

## 12
*Installing the ASCC/Mark I in Cambridge and Transferring It to the Navy*   109

## 13
*Aiken at the Naval Mine Warfare School*   115

## 14
*The Dedication*   121

## 15
*The Aftermath*   131

## 16
*Some Features of Mark I*   147

## 17
*Programming and Staffing, Wartime Operation, and the Implosion Computations*   159

## 18
*The Mystery of the Number 23*   169

## 19
*Tables of Bessel Functions*   177

## 20
*Aiken's Harvard Program in Computer Science*   185

## 21
*Later Relations between Aiken and IBM*   197

## 22
*Aiken at Harvard, 1945–1961*   201

## 23
*Life in the Comp Lab*   215

## 24
*Retirement from Harvard*   227

## 25
*Businessman and Consultant*   231

## 26
*A Summing Up*   237

*Appendixes*

A   *The Harvard News Release*   249

B   *Aiken's Talk at the Dedication*   253

C   *Aiken's Memorandum Describing the Harvard Computation Laboratory*   263

D   *The Stored Program and the Binary Number System*   269

E   *Aiken's Three Later Machines*   275

F   *How Many Computers Are Needed?*   283

G   *The NSF Computer Tree*   295

| | | |
|---|---|---|
| H | *Who Invented the Computer? Was Mark I a Computer?* | 297 |
| I | *The Harvard Computation Laboratory during the 1950s* | 305 |

*Sources*  309
*Index*  325

# *Preface*

The goal of this volume is to present an interesting personality, a scientist and engineer who once was a towering figure in the new world of computers but who is no longer well known. Although *Time* once devoted its cover to a fanciful rendition of one of Howard Aiken's machines, and although there were many other signs of his importance, his true place in the history of the computer is today far from easy to define.

One of the difficulties in assessing Aiken's place in the history of the computer is that most discussions of his era tend to concentrate on genealogies of hardware development. Aiken's computers did not determine the technological path that led in a rather direct line to today's computers, nor did Aiken's work influence the writing or the use of software. Nonetheless, Aiken was one of the major innovators of the computer age, and in the judgement of contemporary witnesses the machine that was developed from his ideas inaugurated the computer age.

Above all, however, Aiken was an interesting person in his own right. His life reads like an American adventure story. Forced to leave school after completing the eighth grade, he eventually completed high school with honors while working full-time at night to support his mother. Even during his undergraduate years, he held a full-time night job to pay for his education and to support his mother. In his thirties, when he had achieved considerable success in the engineering world of public utilities, at an age when most scientists had earned their advanced degrees and were settled in a permanent position, he gave up his career to pursue a Ph.D. in physics. Then, launched on a new life in research in the budding field of electronics, he shifted intellectual gears once again and proposed a giant super-calculator. As constructed by IBM, that machine has generally been considered to have inaugurated the computer age.

Aiken and his team at Harvard built three later machines, each very useful for the purpose for which it had been designed. However, as I have already mentioned, Aiken's ideas did not influence the main lines of development of computer design. His primary achievement was not in computer architecture, which actually followed a different path from his. He did, however, establish the first full academic program (leading to M.A. and Ph.D. degrees)—one that became a model for others. The roster of his students includes many of the next generation of leaders in the computer field. He organized symposia that brought major scientists and engineers in the new field together so that they could learn of one another's progress and problems and thus advance the field by sharing knowledge. He became a super-salesman, preaching the gospel of the computer and its vast potential to universities and businesses all over the United States and Europe. He pioneered the use of computers for billing, and he forced a reluctant Harvard into becoming a computer center. He was one of the first university scientists to understand how government funding could be used to support large-scale projects in the sciences. Clearly such a figure merits attention.

I undertook this volume for several reasons. The first is that, in concert with Henry Tropp, I conducted a two-day oral-history interview with Aiken in 1972, just a few months before Aiken died. This treasure trove of personal history, anecdotes, and insights was so rich that it demanded a book. The second is that I was a party to the planning and organizing of the Pioneer Day celebrations that honored Aiken's memory at the National Computer Conference at Anaheim in 1983. The papers presented at that conference dealt with a number of aspects of Aiken's contributions to the computer. Some of them have now been edited and assembled, together with some further papers about aspects of Aiken's life and career, in another volume published by The MIT Press. While I was editing that volume (titled *Makin' Numbers*), it quickly became apparent that, although the various papers dealt effectively with Aiken's career, the man himself was not being presented to the reader. To provide this missing element, I undertook to produce a "portrait."

Another reason for this book is that in the course of my research on the early history of the computer I found that Aiken's place in history was almost always misunderstood or undervalued. I soon came to realize that it was time for a more balanced view. Yet another is that I knew Aiken as a colleague at Harvard and came to admire his achievement.

As I gathered material for the book, I found that there were many points of contact between Aiken's activities and events in my own education, life, and personal career. Indeed, Aiken's Harvard career intersected not only with my own life but with those of many of my Harvard colleagues, friends, and former classmates. Accordingly, this volume is more personal than the usual biography. Though my own relations with Aiken were always cordial, I have tried not to be biased in his favor but to give an honest portrait. This will be apparent in chapters 13 and 14, where I describe Aiken's stormy relations with Thomas J. Watson and with IBM over the assignment of credit for the "IBM ASCC" or "Harvard Mark I."

Much more space and attention is given to the development, the architecture, and the operational features of Mark I than to the features of Aiken's three later machines, Mark II, Mark III, Mark IV. The reason for this is that Mark I was a pathbreaking creation of enormous historical importance, representing the cutting edge of the art and practice of computing. There is general agreement among many contemporaneous observers and computer scientists that Mark I heralded the dawn of the computer age. By the time Mark II was put into operation, ENIAC had opened the world of electronic computing and EDVAC had announced the stored program and, in a very real sense, defined the modern computer. These steps were so gigantic and revolutionary that in retrospect they may seem to dwarf the magnificent achievement of Mark I. Although Mark III and Mark IV did useful work, they were far from being state-of-the-art machines in the sense that Mark I (and even, to a degree, Mark II) was. Indeed, the very fact that the later machines were needed and were able to do important work is a sign of the growing demand for computing power in the late 1940s and the early 1950s, when computers still tended to be one-of-a-kind machines designed and constructed by the institutions they were to serve. In any event, Mark III and Mark IV represent a technological terminus rather than a node of growth.

In addition to Aiken's machine, I devote considerable attention in this book to his computer science program at Harvard and to the distinguished roster of students who became computer scientists under his tutelage. In the long run, Aiken's innovative service to education in computer science may be his most lasting contribution.

It will be apparent to readers that I had to use information provided by many informants and that I had to piece together many snippets of reports, memoranda, correspondence, newspaper clippings, and other types of archival documents. To provide a sense of authenticity, and also to introduce the reader to some very interesting personalities in their own right, I have, wherever possible, quoted from the sources rather than summarizing.

I am fully aware of the difficulties of understanding all the technical issues of a past age, and I should be grateful indeed to any readers who may have information that corrects any of my statements or that provides further light on issues and events for which I did not have full information or which I have reported inaccurately.

# *Acknowledgments*

This book draws heavily on an interview with Howard Aiken and on discussions with many of Aiken's former students and associates. These include Robert Campbell, Richard Bloch, Grace Hopper, Frederick P. Brooks Jr., Anthony Oettinger, Garrett Birkhoff, John Ladd, James Hooper, James Baker, Ronald King, Edward Mills Purcell, John Palmer, and Cuthbert Hurd. I have also been able to draw on some interviews conducted by Larry Saphire as part of IBM's oral-history program, chiefly those of Benjamin Durfee, Francis Hamilton, Rex Seeber, and Reginald Daly.

I have made extensive use of the Harvard University Archives, a rich repository of Aiken correspondence and other documents, and I am especially grateful to Clark Elliott for his kindness and help in dealing with these materials. I have also been able to draw on the rich resources of the IBM archives, thanks to the kindness of two of IBM's historians, Charles Bashe and Emerson Pugh. The book could never have been completed had I not been privileged to make use of letters and memoranda in the IBM archives, primarily correspondence and documents relating to the ASCC/Mark I. I am grateful to the Harvard University Archives and the IBM archives for permission to use material from their collections.

Tony Oettinger, Aiken's "heir" at Harvard, helped by providing me with information and by making available to me his own rich archival resources; he has also encouraged me in this project from its very inception. I have also had frequent and fruitful conversations concerning Aiken's life and career with Fred Brooks and Dick Bloch, both of whom have been very generous in providing information and interpretations.

I am especially indebted to Maurice Wilkes, who has set me straight on many issues relating to the early days of the computer and has kept me aware of the broader context of computers in the early decades. I also acknowledge the kindness of Grace Hopper, who gave me several interviews from which I gained valuable insights.

I am more than ordinarily grateful to Herbert Grosch, whose severe criticism of an earlier publication of mine has stimulated me to correct an error in computer nomenclature and has prompted me to document more fully certain assertions concerning Aiken's career and his contributions to the computer. I am glad to credit Dr. Grosch for having taught at Columbia, in collaboration with Wallace Eckert, a series of computer courses that seem to have been the first such courses to be offered at an institution of higher learning anywhere in the world. Furthermore, I thank him for helping me to obtain a copy of a rare publication documenting the status of Harvard's program in computer science as of 1949.

I first became acquainted with Greg Welch (who is a co-editor of *Makin' Numbers* and who has been very helpful in a number of ways) while he was an undergraduate in Harvard College. He came to my study in Widener Library with a double request. He wanted permission to write his senior honors thesis on "Aiken and the Mark I," and he wanted to know whether I would serve as his "tutor" and thesis adviser. He was aware of my own research on Aiken and Mark I, and he knew that his work would necessarily anticipate or duplicate my research. I did not wish to discourage him, but I did stress that his research and writing would have to include the very materials I was then studying in preparation for writing this volume. Greg is endowed with a real gift of persuasion, and after a few minutes of eloquent argument he persuaded me to approve his project, to agree to supervise his research, to guide him to resource materials, and to be his tutor. He did agree, however, that his thesis project could not in any way inhibit or interfere with the writing of my own book. Since undergraduate theses are not usually published, we did not foresee any real conflict of interest. On the contrary, we both recognized that his archival research, especially his detailed examination of documents in the Harvard University Archives, could save me time and effort in my own endeavors. Accordingly, I not only made the documents in my personal archive available to him; I also invited him to join me in several interviews (notably with Garrett Birkhoff, Ronald King, and Edward Purcell). I sent him on his own to interview Aiken's second wife, and I arranged for him to visit Aiken's third wife in Fort Lauderdale. Greg's honors thesis is on deposit in the Department of the

History of Science at Harvard University, and portions of it are included in *Makin' Numbers*. I have found it very helpful to discuss certain aspects of Aiken's ideas and career with Greg, and I am most grateful to him for having read the whole of this work in manuscript form.

Paul Ceruzzi, one of the foremost members of the new generation of specialists in the history of the computer, has been of special help. Ceruzzi, currently a senior member of the staff of the National Air and Space Museum, devoted a large portion of his doctoral thesis to Aiken and the ASCC/Mark I; he also discussed Konrad Zuse's contribution to the computer and the significance of Zuse's Z-3 and other innovations. This thesis was later revised and converted into a book, *Reckoners*, that is easily the best source available in print on Aiken and his machines. I am especially grateful to him for giving my manuscript a critical reading and for answering a number of queries.

I owe a special debt to Brian Randell, a friend of many years. Brian not only shared with me his great wisdom and knowledge of computers and their history; he also helped by giving an early version of this book a critical reading. I am particularly grateful to him for contributing information on the stored program.

Michael Mahoney also gave a draft of this book a critical reading and then discussed a number of its features with me. I profited greatly from his comments.

I owe the greatest debt of all to Robert Campbell, who has been my constant adviser almost from the very start of this project. I discussed almost every aspect of the book with him. He read each draft, correcting misinterpretations and errors. I have drawn on his knowledge and experience again and again, and he has been one of my primary sources of information concerning Aiken's career, his machines, and his personality.

I have had the constant cooperation and support of Mary Aiken, Howard's devoted helpmeet of his later years. She has been a continual fount of information, notably about Aiken's early life, and she has helped me to fill in many gaps in information obtained from other sources. She also gave my work the benefit of an early critical reading. One of the fruits of my research has been the bond of friendship with Mary.

In its first stages, the research on which this volume is based was supported by a grant from the Charles Babbage Institute (now the Charles Babbage Foundation). The final research and the writing were supported by a very generous grant from the Richard Lounsbery Foundation. Julia Budenz helped me to complete my information about several aspects of Aiken's career (especially in the pre-Harvard years and the early Harvard years) and acted the part of a prospective reader, alerting me to shortcomings in my presentation that needed attention.

# The Names "ASCC" and "Mark I"

The continued existence of an IBM name (ASCC) and an Aiken-Harvard name (Mark I) is symbolic of the long struggle over who "invented" the giant calculator that bore the two names: IBM's engineers or Howard Aiken. The "official" name assigned by IBM was "IBM Automatic Sequence Controlled Calculator." So far as I have been able to discover, this name was assigned by IBM in the final stages of the calculator's completion. The full name—quite a mouthful—tended to be abbreviated as ASCC, and many IBMers still refer to the machine by that name.

There seems to be no way of learning when Howard Aiken decided to call his first giant calculator Mark I, but very likely it was soon after Aiken agreed to the Navy's request that he build a second machine. Instruments or tools developed by the Navy were usually named for their place in a sequence—Mark I, Mark II, Mark III, and so on. Accordingly, Aiken would have called his second machine Mark II, whereupon the first machine would have become Mark I.

It should be noted, however, that in its officially published manual of operation Mark II is called "the Relay Calculator," and that the manual for Mark III refers to that machine as "a Magnetic Drum Calculator." The progress reports on the design and construction of Mark IV refer to the machine simply as Mark IV, the name used in the title of the manual of operation.

Many difficulties are associated with the IBM name, even in its abbreviated form. For example, there has been some confusion between ASCC and SSEC, the latter standing for Selective Sequence Electronic Calculator. For some it has been difficult to remember the exact initials ASCC and to get them in the correct order, as is evident from Larry Saphire's interview with Frank Hamilton. In the course of this colloquy, it suddenly occurred to both Hamilton and Sapphire that they—like nearly everyone else—had been referring to the machine as Mark I. Recognizing the "error," Saphire remarked that they should not be using that name. The tape then records that they decided to refer to the machine as SCCC—or was it ASSC? Both Hamilton and Saphire had difficulty getting all the letters right and putting them in correct order. I too have always had difficulty getting the initials right; I often have to go back to the full name (Automatic Sequence Controlled Calculator) in order to get the initials ASCC in the proper order and to decide that the abbreviation is not ASSC or even SSCC, which is almost SSEC. Benjamin Durfee and George Daly, in their interview, simply refer to the machine as Mark I. The confusion provides one rationale for referring to the machine by the easier name of Mark I.

A final sanction for using Mark I rather than ASCC is provided by *Father and Son & Co.*, an autobiographical volume written by Thomas J. Watson Jr. Nowhere in that book is Howard Aiken mentioned by name, even in the one short paragraph devoted to the machine. "During the war," according to Watson, "IBM and Harvard University had built a gigantic non-electronic computer called the Mark I," which "consisted of two tons of IBM tabulating machines synchronized on a single axle, like looms in a textile mill." Watson also remarks that "Mark I got a lot of attention as 'Harvard's robot super-brain,' and was used successfully on top-secret war problems." The only part of this recollection that is of concern here is that Watson used the name Mark I and did not even mention that the machine was also called Automatic Sequence Controlled Calculator. Even IBM's official historian, Emerson Pugh, refers to the machine as Mark I in his book *Building IBM*.

# *Howard Aiken: Portrait of a Computer Pioneer*

Howard Aiken in 1961, around the time of his retirement from Harvard.

# 1
## Introduction to a Pioneer

Howard Aiken was a giant of a man in force of will, in originality of mind, and in achievement. Standing six feet four, he towered over most of his students and colleagues. He had a huge dome of a head, piercing eyes, and huge, somewhat satanic eyebrows. He judged others rigorously, and he invited similar judgment. Some colleagues and some former students remained devoted to him for the rest of their lives; others tend to remember only occasions when he was intransigent and difficult.

He related to others in extremes. Almost from the moment he met you, you were either at the top of his scale or at the bottom—there was never a middle ground. But many students, colleagues, and employees came to appreciate him and became unwavering in their affection, no matter to what extent their views on particular issues might diverge or to what degree they might disapprove of some of his behavior. Frederick P. Brooks Jr., the chief engineer in charge of designing and constructing IBM's System/360, remained a friend and admirer long after his graduation from Aiken's program at Harvard. Yet Brooks admitted that, although Aiken was "an admirable man in many, many ways," he was "not uniformly so." Brooks found Aiken's principal "character fault" to be that "he was not unwilling to unleash his full powers on little people—people who were smaller than he was—waiters, airline clerks, and so on."

When Aiken came in contact with an equally strong and assertive character—such as Thomas J. Watson Sr., president of IBM—conflict was inevitable. Harvard's Dean of the Faculty, McGeorge Bundy, is said to have found that the only way to respond to Aiken's table-pounding was to pound the table back at him. Yet only someone with a strong personality and a tough-minded aggressiveness could have made Harvard a center of activity in the new field of computer science at the dawn of the computer age.

The stories told about Aiken's harsher side tend to paint a more colorful picture than those that deal with his kindness to students, his concern that they master their subject, or his patience in explaining difficult problems. Such stories overshadow the fact that Aiken could also be courteous, considerate, and caring, taking real interest in the welfare of his associates' families. Many former students and associates remained fond of him, and at the end of his life he was especially proud of friendly relationships that spanned many decades, some going back to his years as an undergraduate.

Aiken did not necessarily perceive those who disagreed with him as challenging his authority; indeed, in many cases an enjoyable argument formed or sealed a friendship. In conversation he sometimes would emphasize a point by staring at you for a moment in silence; then, blinking and giving his nose a slight twitch, he would let his pince-nez glasses fall; you then knew that you had his undivided attention.

Aiken insisted that his students and associates be properly attired. Always neat and orderly, he expected others to be the same. According to Mary Aiken, his wife in his later years, one of his favorite stories about himself concerned a course he taught after returning to civilian life at the end of World War II. One student showed up for the first lecture looking rather unkempt and sporting a scraggly beard. Aiken took one look at the student and told him to get out. Irate, the student took his case to Harvard's administration. Aiken received a telephone call from the president's office telling him that he was required to admit to his lectures any student who met the course requirements and had paid his tuition. The next day, the bearded student appeared in class and arrogantly took a place in the front row. Aiken looked him squarely in the eye and barked: "OK, you're in. But I want you to know that I'm giving you your final grade right now. It's an E!" An E was Harvard's failing grade, equivalent to the F given at other institutions.

One encounter that occurred in September 1950 in Aiken's Computation Laboratory at Harvard is evidence of both Aiken's powerful personality and his delight in intellectual sparring. Maurice Wilkes, one of the pioneers of the computer, was the principal designer of a machine called EDSAC. Built at Cambridge University, EDSAC was the first digital, electronic, stored-program computer to be put into regular operation. In 1950, when he visited Aiken's laboratory, Wilkes was somewhat junior to Aiken in rank, age, and experience. Yet both men belonged to the tiny group of those who had led computer projects and carried them through to the point where the resulting machines were doing useful work. In a previous encounter, when Aiken had tried to provoke an argument, Wilkes had declined—apparently because, though he greatly respected Aiken's early achievements with Mark I, he did not have the same respect for Aiken's current views on computer design. This time, however, when Aiken tested Wilkes by disparaging the use of binary numerals in computers, Wilkes strongly defended it. Wilkes remembers clearly how Aiken enjoyed this intellectual jousting. Aiken, evidently impressed by Wilkes's intellectual strength and character, invited Wilkes to his home for drinks and dinner.

A quite different aspect of Aiken's personality is revealed by a story told by the late Isaac ("Ike") Auerbach, an important innovator and designer who later became one of the foremost publishers of books on computers. Shortly after World War II, while enrolled at Harvard in a master's degree program, Auerbach took a course in applied mathematics with Aiken. During the spring break he was looking for a job for the summer. Since his home was in Philadelphia, he went to the new firm of Eckert and Mauchly (the Electronic Control Company) in that city. He was promised a position, starting in the coming summer. Thrilled by the prospect and eager to be well prepared, Auerbach told Aiken about the job and asked whether he might stay on for an extra month of additional training, so that by "studying personally with Aiken" he might become "more qualified to work for Eckert and Mauchly than I was at that point of time." According to Auerbach, that was "the last conversation Aiken had with me for four years." In Aiken's view, Auerbach had "gone to work for the enemy." (Those were, Auerbach reported, "effectively his words.") For the rest of the academic year and for some years thereafter, Aiken considered Auerbach "a non-person." During the remainder of the academic year, Aiken would "literally never acknowledge [Auerbach's] presence in the class." Even when Auerbach "would pass him in the street and greet him," Aiken would not acknowledge his presence. When Aiken ran into Auerbach after graduation, he didn't "know" him. But after Auerbach left Eckert and Mauchly for a job at Burroughs, Aiken greeted him warmly, accepting him as "a friend" and "a colleague" and even introducing him as one of his "computing protégés." (Auerbach never learned for certain why or when Aiken had developed so strong a dislike for Eckert and Mauchly. He later supposed that it must have stemmed from Aiken's feeling that his Mark I had been eclipsed by Eckert and Mauchly's ENIAC.)

A quite different anecdote exemplifies Aiken's typical bluntness. John Harr, who eventually became Associate Director of Aiken's Harvard Computation Laboratory, recalls his first encounter with Aiken, early in 1947, when Harr, a graduate student in the Mathematics Department, was trying to get by on his $90-a-month government allowance as a former serviceman. Having seen a job notice for programmers, he rushed to the Computation Laboratory. Aiken asked him "What are you doing now?" Harr replied that he was working toward a master's degree in mathematics. Aiken's reply was immediate and unambiguous: "You might as well get out. I don't want any graduate students." Aiken wanted only individuals for whom the programming job would be a primary concern. Undaunted, Harr returned some time later. Saying nothing about graduate school, he merely indicated that he was trained in mathematics, needed work, and would continue with his education "as time permits." Aiken heard him out and gave him the job. Harr remained on the staff of the Computation Laboratory from 1947 to 1955 and got a Master of Science degree in Applied Science under Aiken in 1951.

Aiken's friends remember his affability and his good sense of humor. The latter shines forth in an anecdote that Aiken told Henry Tropp and me during an interview we conducted in 1973.[1] Describing a visit he had made to the headquarters of the RAND Corporation (a research and development organization set up by the Air Force), Aiken recounted that the director had escorted him on a tour. "He finished the tour," said Aiken, "by sticking his head in the cafeteria" and remarking: "As you can see, we have an extremely able staff. As you can see, for recreation they play Kriegenspiel." "What's that?" I asked Aiken. Hank Tropp explained that it is a variant of chess in which a barrier prevents each player from seeing the other's board and the players don't actually know all the opponent's moves. "That is correct," Aiken interjected. "You can't see your opponent's board. [If] you try to move into a square and it's occupied, you're told you can't do that." But you're not told what piece is there. When Aiken showed up for another visit at RAND three months later, the director "laid on this same tour—the same path, the same place, the same comments, every-

---

1. Tropp, now Professor of Mathematics Emeritus at Humboldt State University in Arcata, California, was the program director of a project—sponsored by the American Federation of Information Processing Societies and the Smithsonian Institution—to interview leading computer pioneers. I served at times as an informal adviser to the project.

thing." Once again, Aiken recalled, the director ended the tour with "the same crack about Kriegenspiel as a recommendation for the brilliance of his staff." By that time, Aiken said, grinning broadly, "I was able to anticipate what might be happening. So, when he told me about the 'Kriegenspiel,' I said, 'Yes, you know, I have an extremely able staff too. They amuse themselves by playing four-dimensional Tic-Tac-Toe.'"

Was Howard Aiken the inventor, or even the first inventor, of the computer? This question raises important problems: exactly what is meant by the words "inventor," "computer," and even "first inventor"? At the very least, Aiken was certainly *an* inventor of one of the early machines from which the computer developed. More significant in the annals of history is the testimony of some witnesses and participants, and also historians and computer scientists, that Aiken's first machine ushered in the computer age. In the widely used *Encyclopedia of Computer Science and Engineering,* Maurice Wilkes declares that "the digital computer age began when the Automatic Sequence Controlled Calculator started working in April 1944." In the same encyclopedia, another article (by E. L. Stoll) begins: "The Harvard Mark I, also called the IBM Automatic Sequence Controlled Calculator . . . marked the beginning of the era of the modern computer." In *Perspectives on the Computer Revolution,* Aiken is given credit for the "real dawn of the computer age," which occurred with "the construction of a machine" that "could control the entire sequence of its calculations, reading in data and instructions at one point and printing results at another."

In 1944, news about the incredible new machine spread so widely throughout the world that Aiken's achievement became known even in Nazi Germany. Konrad Zuse, one of the pioneers in the early development of the computer, learned about Aiken's machine through a notice published (with a photo) in a Swiss newspaper.

In 1959, a "family tree" of the history of the computer was prepared by the National Science Foundation. In the original version of this diagram (reproduced in appendix G below), the main trunk of the tree is labeled HARVARD MARK I. This is a rather incorrect version of history, but it does give some sense of the importance that was assigned to Mark I in the 1950s.

Another sign of the high place accorded to Mark I can be found in a manual on programming published in 1951 by three computer scientists at the Zurich ETH (Polytechnic Institute), Eduard Stiefel,

A. P. Speiser, and H. Rutishauser. In his autobiography, Konrad Zuse relates how he tried to get these three, who were his colleagues at the ETH, to change their minds and acknowledge his priority in the invention of the computer. However, Zuse reports, they persisted in considering Aiken's Mark I "as the first operational computer," and they designated it as such in their book, acknowledged by Zuse as "the first book on computer programming ever."

Aiken's commanding place in the unfolding world of the computer was affirmed in 1964, when he was given the American Federation of Information Processing Societies' inaugural Harry Goode Memorial Award for "outstanding achievement in the field of information processing." In 1965, Harvard University recognized Aiken's enormous contributions by awarding him an honorary degree and naming the computer laboratory he had founded the Howard Hathaway Aiken Computation Laboratory.

In 1980, in one of the first major scholarly studies on the origin of the digital computer, Paul Ceruzzi concluded that "if one were to look for a date from which to mark the dawn of the computer age" it would make sense to choose 7 August 1944—the day the Automatic Sequence Controlled Calculator was publicly announced—as the day "the existence of such machines was first made known to the world."

In 1983, the National Computer Conference, organized by the American Federation of Information Processing Societies, devoted its annual Pioneer Day to a series of meetings honoring Aiken and his machines. The speakers were former students and associates. Aiken received further scholarly recognition in 1985, when a facsimile edition of the Manual of Operation for Mark I was issued. That was followed by a facsimile edition of the proceedings of the Aiken-organized 1947 Symposium on Large-Scale Digital Calculating Machinery.

After the ASCC/Mark I, Aiken went on to build three more machines. He also established the first full university program leading to a master's and a doctorate in the new field of computer science.[2] His students became the pioneers in the new industry. Though at first Aiken believed that computers would serve primarily to solve mathematical problems and to compute mathematical tables, and did not believe that they would also serve the needs of business, later he was

---

2. Courses in computer science did not begin at Harvard, however, until the academic year 1947–48, a year after the inauguration of such courses at Columbia.

in the vanguard of those who showed how to use computers for business purposes.

Like many true pioneers, Aiken's had a multi-faceted career. Collectively, his innovations outweigh in importance the features of any of the machines with which he was associated. The portrait of such an innovative and influential scientist is thus of interest both as the presentation of the life and character of an extraordinary individual and as a source of information on the coming into being of the computer and the dawn of the computer age.

# 2
## *Early Life and Education*

Howard Hathaway Aiken was born in Hoboken, New Jersey, on 8 March 1900. His mother, Margaret Emily Mierisch Aiken, was a child of German immigrants; his father, Daniel H. Aiken, was from a wealthy and well-established Indiana family. Howard was an only child. When he was just entering his teens, he moved, with his parents and his maternal grandparents, to Indianapolis.

Daniel Aiken was addicted to alcohol and, during fits of drunkenness, would physically abuse his wife. During one such episode, young Howard, already large and strong at the age of 12, grabbed a fireplace poker and drove his father out of the house. The family never saw Daniel Aiken again.

Once the father had disappeared, the paternal relatives would have nothing more to do with young Howard or his mother and did not help them financially. Aiken was in the ninth grade when it became his responsibility to support his mother and grandparents. This meant that he would have to leave school and go to work. He got a job installing telephones and began to take correspondence courses. Later in life, he enjoyed telling how he had installed all the phones in the red-light district of Indianapolis.

Apparently, as Aiken told the story of his life and as his wife Mary recounted it to me, there was in his school one of those wonderful teachers who are really concerned about their students' careers and well-being. This teacher, having seen signs of Aiken's intellectual brilliance, especially in mathematics, went to see Mrs. Aiken to plead that her son return to school. Because of the family's pressing financial needs, Mrs. Aiken could not acquiesce. The teacher then found Aiken a night job as an electrician's helper for the Indianapolis Light and Heat Company so he would be able to attend school during the day. The job with the power company, Aiken told Henry Tropp and me when we interviewed him in 1973, was as "a switchboard operator." It was so boring that he took up knitting, making his own wool socks to while away the hours. He was fortunate in being able to get by with only a few hours' sleep.

Insight into Aiken's character as it was manifested at this time can be gained from an exchange of letters between him and an old school friend who was still living in Indianapolis when, in 1944, Aiken's name became well known through newspaper and magazine reports of the ASCC/Mark I. The brief correspondence reveals both a warmth of friendship enduring from those early days and the impression made by the youthful Aiken's scholastic dedication and achievement. Anthony Johnson wrote to Aiken on 11 August 1944:

Dear Friend
Your name, picture and the picture and story of your super-calculating machine has dominated the local papers for several days.

It was only natural that I should feel some pride in being able to tell a few of my present friends and neighbors that at one time I had more than a speaking acquaintance with the inventor. That has been a long time ago and it is apparent you have come a long way since then.

Johnson went on to say that he remembered the days when Aiken was "working so industriously to develop the ability and gain the knowledge" later used in perfecting the calculator. He continued with a mixture of friendly wit and friendly respect by congratulating Aiken "upon an achievement which I know must represent an incalculable (even on your own device) amount of hard work, patience and persistence." In concluding, Johnson mentioned another friend, Elsie, whom he had seen that day and who had asked him to tell Aiken that "the same went for her too." In the midst of all the triumph, controversy, and confusion that marked summer of 1944, Aiken replied on 29 August:

Dear Tony
I was awfully glad to receive your letter, and all the kind things you had to say. Indeed, as I read it, I thought back over the years since we were in school together in Indianapolis, and of the various tricks we used to play. I remember, too, the party your mother had for me just before I left the city for Wisconsin.

I have not been in Indianapolis for years, but hope to make a trip to the mid-west some time after the war. At that time, I shall take the greatest of pleasure in looking you up and talking over the old time things your letter has recalled to mind. Needless to say, if you have the opportunity to come east, it would be a pleasure for me to have you visit with us.

Aiken concluded that he looked forward to seeing Johnson "in the not too distant future," signed as "Your sincere friend," and, in a postscript, sent his regards and thanks to Elsie.

Aiken completed his studies at the Arsenal Technical High School in Indianapolis in 1919, a member of the school's first class to be graduated. His hard work and his good academic record won him the friendship and interest of a school official. Aiken told Tropp and me of the importance of the support he received from "the Superintendent of Public Instruction, whose name was Milo Stewart." Because Aiken was short of credits, Stewart set up a special examination for him, Aiken said, "so that I could get some credits and get out of high school and get off of these rather bitter twelve-hour nights." Stewart also helped Aiken by writing about possible employment to every midwestern public utility that had a base of operations in a university town. As a result, Aiken told us, he was offered a job with the Madison Gas and Electric Company. That, he said, was "the reason I went to the University of Wisconsin."

Aiken and his mother moved to Madison, where he enrolled in a program in electrical engineering at the university. To support his mother and himself, he worked on the night shift as a "watch engineer" and apparently as a telephone operator for the Madison Gas and Electric Company while attending college during the day. The four-to-midnight shift was "much easier" than the twelve-hour shift at Indianapolis. During his junior year, Aiken later recorded, he was also a professor's assistant in physics. In 1923 he was awarded a B.S. degree in electrical engineering.

During his college years, Aiken had set up his mother in her own apartment; he had lived with two or three roommates. One of them, Thomas Leonard, went on to study medicine and later became a member of the faculty of the University of Wisconsin School of Medicine. Leonard achieved great success in clinical teaching and in his medical specialty, gynecology. He and his wife, Myrtle, remained friends with Aiken for the rest of Aiken's life. In a letter of condolence to Mary Aiken after her husband's death, Leonard recalled how poor he and Aiken had been as undergraduates. When we "each started on our preparation for life," he recalled, we were "poorer than anybody." But we "never permitted this to even remotely deter us from continuing to achieve what we set out to do." "That," he supposed "is one of those things which helped us to smile at the difficulties which might have discouraged others." "For me," he added, "and I know that I can speak for Howard, our individual contributions have been of some significance. We have left the world a little better for our having been here. We can look back and be thankful that these capabilities have been given us, and we can be happy that the opportunities arose through which we can utilize them."

In college, as in high school, Aiken won the respect of his teachers. Edward Bennett was clearly of particular importance. Born in 1876 and therefore 43 years of age when Aiken was a 19-year-old freshman, Bennett had received his degree in electrical engineering from Western University of Pennsylvania in 1897. During his early working years, as a research engineer for Westinghouse, he had been active in the then-new field of electrical engineering. In 1909, after holding positions such as that of head electrician in various companies, Bennett joined the faculty of the University of Wisconsin as associate professor of engineering. He attained the rank of professor in 1913 and became chairman of the department in 1918, the year before Aiken entered the university. Twenty-five years after this matriculation, in a letter inviting Bennett to the dedication of the ASCC/Mark I, Aiken wrote to his former professor: "I sincerely hope it will be possible for you to be with us, for in a large part, successful completion of this machine was due to the careful preparation which I had as your student." He signed the letter "Respectfully yours."

In a lecture delivered in 1955 in Sweden and in Germany, Aiken referred to the importance of Bennett's teaching in the shaping of his thinking process. He recalled how Bennett had taught him how "the development of any new body of knowledge" went through four stages. The first was *observation,* when "the investigator knew so little about the subject" that "the only useful thing he could do was to make observations in nature, observe new facts." The second was *classification,* when the observer had enough facts so that he could "separate" them "and place them under major headings" and then "rank these facts in the order of their importance." The third stage, *deduction,* was "when a new science is born." It was now possible "to deduce new facts from old." As a result, it was no longer "necessary to go to nature to make an observation each time a new piece of information was required." Finally, there was the *methodological* stage. Now "the demonstrations of the masters were accepted without question and a super-abundance of words stifled thought." Aiken showed his audience how Bennett's schemata was exemplified in the history of the computer.

A few years later, making comparisons with the kind of academic program young men could follow at Harvard, Aiken expressed some reservations about his own undergraduate education. Writing to Walter Beckjord,[1] a friend and associate from the years in Madison, he referred to the "technological drudgery" of certain courses that had been imposed upon him as a student. He was "greatly pleased" that the courses being taken by Beckjord's son as a Harvard freshman included physics, mathematics, English, and Greek. "I only wish," he continued, "that as an undergraduate I had been so well advised, rather than to have wasted my time on mechanical drawing, machine shop, and other tommy-rot required of a freshman in a conventional engineering school."

Because of discrepancies in the record, it is difficult to determine the precise chronology of Aiken's career between 1923, when he was graduated from the University of Wisconsin, and 1932, when he matriculated at the University of Chicago as a graduate student in physics. The most probable sequence of places and dates situates him at Madison Gas and Electric from 1923 to 1927, at the Westinghouse Electric Manufacturing Company in Chicago from 1927 to 1931, and at the Line Material Company in Detroit in 1931 and 1932.

It is certain that, after his graduation from Wisconsin, Aiken stayed with the company for which he had worked while studying for his degree. His new position, as he described it later in his curriculum vitae, was that of "electrical engineer responsible for design and reconstruction of the company's electric generating station." During the 1973 interview, however, he told Tropp and me that he "became

---

1. Born in St. Paul in 1888, Walter Clarence Beckjord received his B.S. in electrical engineering from the University of Minnesota in 1909 and then had a long career, first as an engineer and then as an executive, with utility companies in St. Paul, Madison, Boston, New York, and Cincinnati. Having moved to Madison from St. Paul in 1916, he became general superintendent of the Madison Gas and Electric Company, with which Aiken was connected from 1919 until 1927. During this period Beckjord was associated with Madison Gas and Electric's parent company, American Light and Traction, of which he eventually became vice-president and chief engineer. In the summer of 1944, he was in New York serving as vice-president, general manager, and director of the Columbia Gas and Electric Corporation and president of the Columbia Engineering Corporation.

Chief Engineer of that company." "I was," he said, "promoted from switchboard operator to Chief Engineer overnight when I got my . . . degree."[2]

Aiken gave further details concerning his career in a letter to Warren Weaver dated 5 September 1940. Weaver, a member of the faculty of mathematics while Aiken was an undergraduate at Wisconsin, had actually been his teacher in a calculus course. Later, during World War II, Aiken was in contact with Weaver, then Director of the Division of Natural Sciences of the Rockefeller Foundation and Chief of the Applied Mathematics Panel of the Office of Scientific Research and Development. "During this time," Aiken wrote to Weaver of his years as engineer at Madison Gas and Electric, "I completely redesigned and rebuilt the electrical equipment in the company's power house, supervised the construction of a three million cubic foot gas holder, and made many other improvements in the company's gas works."

Further characterization of Aiken's professional achievements of that period appears in a letter from Walter Beckjord dated 16 August 1944: "It seems quite a long time ago since we were associated in the utility business out in Madison and I recall very well the very broad-minded way in which you approached the problems of outdoor substations for 2200 volt switch gear and, as I remember it, the substations in Madison were built out of doors as a result of your studies up in St. Paul." This testimony is especially important because Beckjord was a kind of mentor to Aiken as well as a professional associate and a longtime friend, and because he was also an accomplished and successful engineer and businessman. Beckjord's great ability is suggested in Aiken's reply of 29 August 1944: "Thank you for your letter of August 16, and for the thoughts it brought to mind as you summed up experiences since Madison where I received my first opportunity due entirely to your inventions at the time when the switch required rebuilding."

Other aspects of Aiken's life in Madison are evident in an exchange of letters, also in 1944, between Aiken and his friend Thomas R. ("Bob") Halbrook. A picture emerges of warm friendship on many levels. Halbrook wrote to Aiken on 20 August:

---

2. At that, point Aiken started to laugh so loudly at his own remark that the final words are not fully distinct. Our transcript of the tapes reads "when I got my master's degree," but Aiken never did receive a master's degree from the University of Wisconsin. Of course, it is possible that Aiken had a lapse of memory and did think he had received a master's degree. Most likely, therefore, what he said (or meant to say) was "when I got my bachelor's degree."

I have thought of our days together—and the Ice box locks, lamp shade machines—BTU's squeezed out of the air, bridge and aluminum atlases, of "Duncan Phyfe" chairs around a rickety card table. . . . Are there any little "Aikens" taking dishes out in the yard and washing them with the hose and other equally creative methods of doing chores as I believe you did? . . . I was completely befuddled by the task of how a lowly Private in the Army should go about addressing a Com. USNR, so I just skipped that part. My respect for you was established long before the acquisition of the braid, and shall continue after it has been laid aside.

Aiken's reply, dated 29 August, reveals further elements of the friendship: "I remember the days your letter suggests in which we talked of art and science, inventions and people, and of ourselves and our future. I have often thought of you since I left Wisconsin."

Aiken's friendships with Jo Walker Humphrey and her husband, Rowland Stokes, probably also date back to his years in Madison (1919–1927). In correspondence carried on between 21 August and 28 November 1944, Mrs. Stokes wrote: "Rowland always said you were a mathematical genius." She concluded one letter with "congratulation, happiness and health to the only genius I ever knew." Rowland Stokes was obviously one of Aiken's close friends, perhaps a fellow student and surely a colleague in electrical engineering, who also knew and cared about Aiken's mother. Having learned from Jo Walker Humphrey that Stokes was associated with a research project at the California Institute of Technology, Aiken wrote to him on 30 August 1944. The bantering tone reveals one aspect of their friendship:

For brevity, this isn't too bad a summary. See if you can't knock out one of about the same character, and send it along. Also, let me know when you come east, because I still have a number of debates, which have not been brought to a satisfactory conclusion. In short, I still think you're bats.

Stokes's reply, begun on 24 August and sent in a letter dated 15 September, introduces another note:

I knew you would pull something spectacular some day so I would find out where you are. I have bitterly regretted losing track of you, and have for years asked every engineer from the east with whom I have come in contact if they had run into you.

Stokes also explained that he was suffering from tuberculosis. The response this elicited suggests deep commitment and caring on Aiken's part:

I was happy to receive your long letter, and disturbed to find that you have a private fight on your hands with that well-known adversary, t.b. He is a formidable antagonist but fortunately one which can be finally downed provided only that he be fought carefully, courageously, and continuously until he is thoroughly licked. I have confidence in you, and am prepared to predict the outcome of your encounter. I trust you have the best of medical care. If there is any question in your mind on the care you have, let me know, that I may consult the faculty of the medical school and find out who in your locality should be entrusted with your care. If you find it difficult to write, I suggest that your wife write me, and keep me informed as to your progress.

Robert Campbell, Aiken's deputy during the final stage in the completion of the ASCC/Mark I, recalls an amusing story that Aiken used to tell about the days when he was working for Madison Gas and Electric. There was a time when, every afternoon shortly after 3 o'clock, there would be a sudden and inexplicable failure in the power lines. For some unaccountable reason, a "short" somewhere along the lines would cause a power outage. Aiken's boss wanted to know why Aiken couldn't find the cause of this problem. After all, Aiken was supposed to be a trained engineer. Aiken was at his wit's end until he received a telephone call from a schoolteacher in an outlying suburb of Madison. Near the school, she reported, an old barbed wire fence had been discarded and cut up. One boy, on his way home from school, would pick up a piece of barbed wire and throw it up at the power lines until he made a good contact and was rewarded by the production of a huge spark. The mystery was solved.

In 1927 Aiken resigned from Madison Gas and Electric "to enter," as he explained to Warren Weaver in 1940, "the Central Station Division of the Westinghouse Electrical and Manufacturing Company." Of his work there he wrote in his curriculum vitae that he was a "general engineer engaged in application of the company's products to designing of electric generating stations." This change in employment took him to Chicago, where he was friend and associate of a certain John Robinson, whose connection with Aiken and whose death is noted by Walter Beckjord in his letter of 16 August 1944. To this reminiscence Aiken responded as follows in a letter dated 29 August: "I often think of Robbie whom I have always regarded as both your friend and mine. Fortunately, I talked to him on the long-distance telephone just a few days before he died. He had been ill but was convinced that he was on the road to recovery." Of that period Beckjord wrote: "Even at that time you were keen to continue your studies in mathematics and discussed with me one day the possibility of going back to school to continue your studies. You certainly were very wise and farsighted in following your bent. . . ."

Before going to graduate school, however, Aiken made another change in employment, resigning from Westinghouse in 1931 "to become," as he later told Warren Weaver, "district manager of the Line Material Company in Detroit, Michigan." "During this period," he continued, "I increased my income, but got too far away from technical and scientific employment. Becoming dissatisfied, I again resigned in 1932 and entered upon further research work and study." With the encouragement of his old grade school teacher, he enrolled in graduate school.

He went first to the University of Chicago, where he matriculated as a graduate student in physics in the autumn of 1932 and where he remained for only two quarters of the regular academic year. Chicago had achieved great renown as a center of advanced research and training in experimental physics, an area built up by A. A. Michelson and Robert Andrews Millikan. Michelson had been professor and head of the department of physics from 1892 to 1929, and Millikan had been on the physics staff from 1896 to 1921. But in 1932, when Aiken arrived, Millikan had been at the California Institute of Technology for 11 years and Michelson was dead. The university was also undergoing a tremendous structural upheaval under the helm of its new president, Robert Maynard Hutchins.

Aiken didn't like the Chicago program; in fact, he later wrote that he had found Chicago "a lousy institution." He discovered, as he put it during the 1973 interview, "that the faculty were bootlegging grades at the behest of the newly appointed president, Mr. Hutchins." And so, "after two quarters I decided to go to Harvard." Aiken went on to tell us that he "had an arrangement by which when I got a doctorate, I would go back to the University of Wisconsin, where I had an offer as an assistant professor of electrical engineering." He never did take up that offer, however, and when he "returned to Wisconsin to go to commencement, when they gave me an honorary degree" he told the president "that I guessed the best I could say was that I was still bucking for that job as assistant professor." Aiken recalled with laughter that "before I got up from the luncheon table, a contract was put in front of me." According to Aiken's curriculum vitae, he received a medal for achievement from the University of Wisconsin, not an honorary degree.

When I asked Aiken how he had happened to choose Harvard, he replied that one of his professors—the one "who impressed me most at Chicago"—had "told me he thought I'd like it at Harvard. I got some catalogues and found out who the men were who were there, and decided it looked pretty good." After spending part of a year at the University of Chicago, Aiken entered the Graduate School of Arts and Sciences at Harvard, thereby starting an academic affiliation that, save for Aiken's wartime leave to serve in the Navy, continued until his retirement in 1961.

Throughout these years and afterwards, Aiken always maintained close relations with his mother. During the period in Wisconsin, as his friend Halbrook recalled in August of 1944, friendly pleasures included being seated at a card table, where "one would find, you, your mother, T. L. [apparently Thomas Leonard], sometimes the L. Hardys, and myself." In the same month, Aiken wrote to Jo Walker Humphrey: "Mother is living in Cambridge, and has just recovered from a very serious operation. Her address is 8 Plympton Street, Apartment 43." When he entered Harvard as a graduate student in 1933, Howard Aiken's address was 820 Massachusetts Avenue, where a YMCA now stands, but from 1934 (if not earlier) he and probably his mother lived in an apartment house at 8 Plympton Street. (The building is still standing; on the first floor is the Grolier Book Shop, one of the few bookstores in the world devoted exclusively to poetry.)

During one summer, probably in 1936 or 1937, Aiken and a fellow graduate student in physics, Bert Little,[3] rented a summer house in Woods Hole, where they and Aiken's mother spent a month. Little was one of Aiken's closest friends during his graduate years at Harvard. Later on, after Little was married, he and his wife often visited Mrs. Aiken in her apartment. Mary Aiken, his wife during his final years, also recalls visiting Mrs. Aiken in the Plympton Street apartment.

In later years, after World War II, when Aiken was married and lived in the suburbs, he continued to keep his mother in the apartment at 8 Plympton Street. Five days a week, as regularly as clockwork,

3. Elbert ("Bert") Payson Little, who later became one of Aiken's associates, had a rather unusual career. After being graduated from Harvard College in 1934, he entered Harvard University as a graduate student in physics, serving as an assistant in Theodore Lyman's course in optics. In 1938 he left Harvard to become an instructor in science at Philips Exeter Academy, taking time off in 1941 to complete his Ph.D. In 1948 he accepted a post as physicist for the Air Force, with the primary assignment of developing a large-scale computing center at Wright Field in Ohio. For two years he was assigned to work with Aiken at the Harvard Computation Lab, serving as liaison between Aiken and his staff and the Air Force; he also was a sort of dean to the Air Force officers who were studying at Harvard.

Professor Aiken could be seen striding across Harvard Yard at a few minutes before noon on his way to his mother's apartment for lunch, which was a hot meal always preceded by a martini or two. This routine was broken only when Aiken had an occasional academic luncheon, was out of town, or had visitors who stayed later than noon. On those occasions, Aiken's secretary, Betty Jennings, would telephone his mother to tell her that her son would not be coming.

Aiken's clockwork precision about lunch evidently continued until the end of his life. In 1973, while Hank Tropp and I were interviewing him at his home, Aiken jumped up and said "How about a bit of lunch, fellows?" I looked at my watch and noted that it read exactly noon. We got into Aiken's car. He explained en route that he was taking us to a "special restaurant." He did not tell us in what sense the restaurant was special, but when we got there and were seated at our table Aiken turned to us and said "How about a martini?" Tropp, whom I had briefed about Aiken's regularity concerning lunch, could not restrain his grin as he looked at me. Aiken stared at me coldly, evidently aware that I had told Tropp about his habits. Aiken then explained that the reason he liked to have lunch at that particular restaurant was that it stocked a certain gin that he favored. When we had finished our drinks, Aiken turned and asked "How about another?"

An interesting aspect of Aiken's daily visits that I learned only during the course of writing this book reveals the warm and humane side of his personality. Betty Jennings informed me that a primary reason why Aiken encouraged his mother to cook a warm lunch for him was to make certain that she would have at least one full meal a day. Evidently, Aiken was aware that older people living alone tend not to bother with the labor of preparing a full meal just for themselves. Aiken's solicitous care of his mother shows a quality of his makeup that was not always apparent on the surface and that probably would have surprised many contemporaries who knew only the stern and unyielding public figure with whom relations could be so difficult.

# 3
## A Harvard Graduate Student

When Aiken began his graduate studies at Harvard, in the autumn of 1933, he was 33, some 10 years older than most of his peers.[1] At Harvard in those days there were three academic groups interested in electricity. One was the Electrical Engineering faculty, located in Pierce Hall (the Engineering School). A second included the members of the Physics Department who were concerned with classical electromagnetic theory and electromagnetism; they were located in the Jefferson Physics Laboratory and in the Research Laboratory of Physics (now the Lyman Laboratory). Yet a third, centered around E. L. Chaffee in the Cruft Laboratory, had the official name Communication Engineering. This third group—often known informally as the "Cruft group"—had four stars: E. L. Chaffee (a specialist in the theory of vacuum tubes and circuits), Ronald P. W. King (whose field was applied electromagnetic theory, transmission of radio waves, and theory of antennas), Harry Rowe Mimno (an electronics expert who was interested in radio transmission, reflection from the ionosphere, and aspects of ultrahigh-frequency radio waves), and Theodore V. Hunt (whose specialty was acoustics). The "grand old man" of this electronics group, then reaching retirement, was George Washington Pierce, who had been a pioneer in many aspects of the physics of radio and of high-frequency alternating currents and who was a major figure in applied magnetostriction. Nominally members of the Physics Department, members of

---

1. This was unusual but by no means unheard of at Harvard in those days. In that same fall of 1933, when I entered Harvard as a freshman, Charlie Smith, the stepfather of my classmate William Vick Smith, was enrolled as a graduate student. Charlie Smith—an inventor whose early patents provided the foundation for the Raytheon Corporation—was working for a belated Ph.D. in physics. He was some 20 years older than Aiken.

the semi-autonomous Cruft group had an anomalous position in that the foci of their interests lay in what was then generally considered an applied area of the spectrum of knowledge, although their research and teaching were directed more toward basic theory than toward practice. Chaffee, the group's leader, was a powerful figure in the university and commanded the respect of industry and of his peers in the general area of vacuum-tube theory and electronics. A patrician in the old Harvard tradition, Chaffee was gentle and affable in mien yet skilled in matters of academic power and politics—an able administrator, firm in his principles, and wise in his decisions. During World War II, Chaffee organized and directed Harvard's large and complex teaching operation for Army and Navy officers—a course on physics and mathematics, including some knowledge of the principles of electricity, magnetism, and electronics they needed to become specialists in radar (in which a technical followup course was available at MIT).

Although the Research Laboratory of Physics and the Cruft Laboratory are at right angles to each other, their floors are on the same level and there is no wall or door separating them. There is also easy access between the Research Laboratory of Physics and the old Jefferson Physical Laboratory. Just as there was an easy transition between the office and research space of the faculty of physics (in "Jeff" and the "Research Lab") and the faculty of Communication Engineering (in Cruft), transition from one area to another was easy for students. Indeed, many a course could be taken either as a "physics" course or as a course in "engineering sciences."

Chaffee's character and scientific expertise commanded Aiken's respect and affection, and he became the guiding figure, sponsor, and protector of Aiken's Harvard career. Ronald King, who came to Harvard as a faculty member at about the same time that Aiken entered graduate school, recalls that Aiken did not get on well with the Physics Department. King no longer remembers, nor have I been able to find out, the cause of the differences. It may be conjectured, however, that Aiken—a tall, intelligent, somewhat arrogant, assertive, older graduate student—did not easily slide back into pupil status. In any event, King recalls, Aiken was practically "kicked out" of the classical physics program. He was rescued by Chaffee, who evidently recognized Aiken's ability and accepted him as one of his graduate students.

As a thesis topic, Chaffee assigned Aiken an aspect of space-charge conduction.[2] Aiken was allotted a small space, to be used as a combined office and research laboratory, in the basement of the Research Laboratory of Physics (or Physics Research Lab), the name then generally used for what has since been renamed the Lyman Laboratory. Edward Mills Purcell, who was a fellow graduate student in the Cruft Lab group, remembers that Aiken took readily to Chaffee's suggestion of a thesis topic. Aiken received his A.M. in physics in June 1937 and completed his thesis (Theory of Space Charge Conduction) in the autumn of 1938. He was awarded his Ph.D. in physics at the winter commencement in February 1939.

Aiken began teaching at Harvard in the academic year 1935–36, before he received his master's degree; his title was Instructor in Physics and Communication Engineering. In this first year, he was a teaching assistant in four courses during each semester: Physics 3a and 3b, Physics 21 and 22, Physics 24a and 24b, and Physics 26 and 28. The annual report of the department listed the subject of his research as "Theoretical Investigation of Space Charge Laws."

In those days the teaching load for a course assistant was far more onerous than it is now. For example, in Physics 3a (Electricity, Magnetism, and Direct-Current Electrical Measurement), taught in the fall term by Associate Professor Harry Mimno, Aiken not only ran several lab sections but also had the assignment of explaining difficult problems to students. He performed the same functions in the spring term for Physics 3b (Electromagnetism and Elementary Principles of Alternating Currents), taught by E. L. Chaffee.

In 1936–37 (and again in 1937–38 and 1938–39), Aiken had additional assignments, one as an instructor in Physics and Communication Engineering and one as a tutor in the Division of Physical Sciences.[3] He continued as a course assistant until 1939, when he was awarded his Ph.D. During 1938–39, however, Aiken gave a course of his own for the first time: Physics 29a (Applied Mathematics).

2. In those days, it was the custom for some professors to assign research problems for graduate students to pursue, rather than allow students to select their own research topics.

3. The title "tutor" refers to a special feature of Harvard undergraduate education: almost every department provided individual or group tutorial work in additional to the regular instruction in formal classes. In most departments these tutorials were one-on-one sessions between an undergraduate and a faculty member, leading to an honors thesis in the senior year.

On receiving the Ph.D., Aiken was appointed Faculty Instructor in Physics and Communication Engineering, effective as of September 1939. He no longer had to be a course assistant. Instead, he was one of the lecturers in Physics 3a, now listed as given by "Associate Professor Mimno, Assistant Professor N. H. Black, Dr. Aiken, and Assistants." Physics 3a, given in the fall semester, was followed in the spring by Physics 3b, now listed as given by "Professor Chaffee, Dr. Aiken, and Assistants." Aiken was also one of the two lecturers in Physics 28a (Electron Physics—Electron Emission, Space Charge, Contact Potentials, Fluctuation Phenomena, and Vacuum Tubes), listed as given by "Professor Chaffee and Dr. Aiken." In 1939 there were 25 students in Physics 3a, 19 in Physics 3b, 6 in Physics 28a, and 30 in Aiken's own course, Physics 29a (Applied Mathematics). Aiken summed up his teaching assignments in a September 1940 letter to Warren Weaver: "I give a course in Applied Mathematics, and am associated with Professor E. L. Chaffee in a course on Electron Physics."

Aiken's curious title of Faculty Instructor was introduced at Harvard in 1939 as a replacement for the standard college and university category of Assistant Professor. The occasion for the change was Harvard's revamping of its rules of tenure and promotion. In the old system, some individuals had been kept on indefinitely at the rank of Assistant Professor without ever achieving permanent tenure. When the title of Assistant Professor was eliminated, the title of Faculty Instructor was introduced. A person with the rank of Faculty Instructor held a five-year appointment of the "up or out" kind. During the fourth year, it would be determined whether he would gain permanent tenure as an associate professor or be "terminated."

Aiken recalled, during our interview with him, how ridiculous he felt when he was promoted to the rank of a Faculty Instructor—as if, he said, his job was "to instruct the faculty." At other universities, however, the rank of Instructor was usually given to someone who had just obtained the Ph.D. and who held an annual appointment, often renewable only for three years. Thus, a Harvard Faculty Instructor would appear to the rest of the world to have held a position of lower rank than an Assistant Professor, or to have not "made it" to the rank of Assistant Professor. The misunderstanding concerning the significance of Aiken's title is reflected in the fact that many published accounts describe Aiken as having been only a "Harvard Instructor" at the time that he proposed the ASCC/Harvard Mark I. (Because of this kind of misunderstanding, Harvard eventually abandoned the rank of Faculty Instructor and adopted the more customary title Assistant Professor.)

As an extra assignment, Aiken gave a two-semester survey of physics at Radcliffe College, Harvard's "sister" institution. In those days of gender separation, Radcliffe students were not allowed to take elementary courses at Harvard; they were admitted only to "upper-level" courses. As a result, Harvard faculty members (usually of junior rank) were given the opportunity to teach a course for the Radcliffe women, for which they were paid a small supplement to their regular Harvard salary. One of Aiken's Radcliffe students is said to have recalled later that Aiken once spoke in class about his dream machine, his super-calculator.

We can gain a good picture of Aiken's quality of mind and of his relations with undergraduates in his courses from the recollections of one of the students in Aiken's course on electricity and magnetism: James R. Hooper, a college classmate of mine who later had a distinguished career at Case Western Reserve University. Hooper recalls that in one of the lab problems, the students measured and plotted a hysteresis curve in $B$-$H$ coordinates and were "given an instrument for measuring the enclosed area," which was "a measure of the energy dissipated in one cycle around the loop." This instrument, a planimeter, was a kind of "hinged two-legged device." One end "remained fixed while the other traced the perimeter of the closed curve." A small wheel rolling over the surface during this motion was linked to "some measure of the changing angle at the hinge point," and the results were registered on a little dial. In order to interpret the result, the planimeter was calibrated by measuring the area of a square in the same $B$-$H$ coordinates. One day, as the lab session was "drawing to a close," Hooper "became curious about how this miraculously intricate instrument worked" and asked Aiken to explain its operation. Aiken had no more idea than Hooper about how the planimeter was able to convert the motion of the wheel around the curve into an area. According to Hooper, Aiken "sat down with a sheet of paper" and, "after a few false starts," worked out the mode of operation. What particularly impressed Hooper was not so much that Aiken could discover how the planimeter produced its results as the fact that Aiken solved the problem by beginning with the simple basics and logically deducing the consequences. It was "the first time," Hooper recalled, that he had "ever seen anyone actually *do* a piece of physics from scratch" as distinguished from "textbooks and lecture demonstrations." Aiken worked out the problem "in terms of differential distances and angles." Some 60 years after the event, Hooper still has a vivid recollection of—"also for the first time"—seeing "the Calculus lifted from Osgood's dusty pages[4] and put to a practical innovative use." Besides showing that Aiken approached problems in a logical manner, and besides serving as evidence of the friendly relations Aiken established with interested students, Hooper's recollections are important because they concern a commonly used analog calculator of the pre-computer era and the way in which it converted angular displacements into areas. This incident may be taken as evidence that Aiken had encountered calculating devices years before he undertook to design a giant and complex calculating machine.

Jim Hooper so enjoyed his contacts with Howard Aiken that he continued to meet informally with him "on a fairly regular basis" throughout his junior and senior years. He always found Aiken "warm and encouraging." Upon entering graduate school in 1937, Hooper planned eventually to do a thesis on some aspect of space charge in vacuum tubes under Aiken's direction and so asked to have Aiken as his advisor. Hooper never wrote his thesis under Aiken, however. Upon marrying, in 1939, he took a teaching job to support himself and his new wife. He started out at Williams College, then went to Union College, returning to Harvard after Pearl Harbor when E. L. Chaffee invited him to return to join the teaching staff of the special course in electronics designed for pre-radar Army and Navy officers. When the war was over, Hooper reentered Harvard and completed his Ph.D. work under Chaffee rather than Aiken, since he was still interested in vacuum-tube circuits and Aiken was now wholly concerned with computers and applied mathematics.[5]

Hooper recalls a day when Aiken had just returned from a visit to IBM (which must have taken place in the spring of 1940, a few months before Hooper left for Williamstown). With enormous pride and delight, Aiken showed Hooper something that looked like "an enlarged roll of stiff toilet paper with holes punched in it." This was plainly a roll of the card stock used by IBM, punched with holes to indicate the instructions for the giant calculating machine then under construction at IBM's facility at Endicott, New York. For Aiken the possession of an

---

4. Hooper and I learned the calculus from William Fogg Osgood's *Introduction to the Calculus* and *Advanced Calculus*.

5. Hooper's thesis was titled Cavity Mode Coupling in an Electronic Space-Charge in a Magnetic Field.

actual punched tape must have been a thrilling event—a foretaste of the victory to be achieved when the machine would be completed and put into operation.

Hooper also recalls a kind of analog device that Aiken constructed sometime around 1937. Aiken was still working in a small research lab and office in the basement of Harvard's Physics Research Laboratory. According to Hooper, Aiken was busily engaged in research on space-charge problems in vacuum tubes. In order to study "electron trajectories in a triode," Aiken constructed a hydraulic analogue. He obtained a shallow tank about 4 feet wide and 6 feet long and placed it on a table. Across the tank he "stretched a rubber sheet membrane," which was elevated to a height of about a foot at one end, corresponding to the cathode of the vacuum tube. A pump was continuously circulating water so that "a sheet of water flowed down the membrane," representing "electrons flowing to the anode." There was a set of equally spaced vertical rods set into the tank, which "poked up from below in a row across the sheet" in such a way as to "represent the grid wires." These rods distorted the membrane, "creating cusps with intervening valleys," through which "the stream of water had to make its way." These peaks or cusps would thus act on the water flow in a manner analogous to the effect of grid wires in a vacuum tube on the flow of current. Aiken would introduce "a narrow stream of colored dye into the water" at "selected points along the cathode end," and would observe "how the dye path flowed past the grid." The flow of the dye through the "grid" must have been an impressive sight. According to Hooper, though, the device had many problems. One was the difficulty of finding appropriate "wetting agents." Another was how to control the "turbulence in the water flow." Hooper doubts that Aiken ever made serious or continued use of this analogue device; he noted that the equipment "looked pretty dusty" in 1940. Hooper recalls that when Aiken showed him the above-mentioned roll of tape, "his hydraulic analogue table was gathering dust" in a corner of a room in the basement of the laboratory. The existence of this device is of interest primarily as testimony to Aiken's ability to visualize abstract mathematical or physical situations in terms of operating physical analogues.[6]

As one of Aiken's graduate students, Hooper read Aiken's thesis. I asked him what he most remembered about the thesis and whether

6. The construction and operation of this device evidently made a deep impression on Aiken's contemporaries. Hooper is not the only one to have a

he ever discussed any aspects of it with Aiken. What remained prominent in his memory, Hooper told me, was Aiken's pride and delight in having been able to master a difficult differential equation. "In those pre-computer days," Hooper recalls, "most differential equations could be handled only by approximate methods." "It was a real achievement," he continued, "to develop a classical solution in closed form." Hooper remembers that "Chaffee was immensely impressed and seemed to bask in the reflected glory of one of his graduate students." Chaffee "even mentioned it in classroom on one occasion."

Hooper remembers that "when Howard was displaying his punched tape and explaining what he expected of his new computer-to-be" he said that his "first goal would be to publish a complete set of Bessel and related functions." As is well known, one of the early products of the new machine was a set of tables of Bessel functions.

As a successful teacher and a brilliant young scientist, Aiken advanced in rank at Harvard. As of 1 September 1941, he was given tenure as Associate Professor of Applied Mathematics. In the official announcement of his promotion it was noted that, because he had "reported for active duty as Lieutenant Commander, Engineering, Special Service, US Naval Reserve," he had "not yet served actively in this position."

When Aiken achieved tenure at Harvard, his giant calculator was still in the final stages of construction by IBM. Harvard's president James Bryant Conant and some senior members of the faculty had already been in close contact with the progress of the machine and had come to know of Aiken's vision and the way his dreams were being brought to fulfillment by IBM.

With evident relish, Aiken told Tropp and me that as a junior faculty member he had been called into Conant's office and told that if he persisted in devoting all his time and energy to computing rather than to research in electronics there would be no future for him at Harvard. I do not know what changed Conant's mind about the promotion; perhaps it was the recognition of the importance of the calculator project and the fact that it was actually being brought to fruition by IBM. Another factor in Aiken's being tenured was the availability of a

---

strong recollection of it. When Aiken's fellow graduate student Ed Purcell was discussing Aiken's early Harvard years, one feat of Aiken's that was still vivid in his memory after almost half a century was the hydraulic analogue of the problem in electron flow.

special bequest by Harvard Professor Arthur Edwin Kennelly that enabled Conant to create a wholly new position in what had become Aiken's new area of primary research interest, as reflected in his formal title. Aiken's new title indicated official recognition that Aiken's specialty was no longer electron physics or communication engineering but applied mathematics—the area Kennelly had specified in his bequest.

In Harvard's annual report for the academic year 1941–42, which contains the official announcement of Aiken's promotion, it is noted that Aiken had been appointed to "a new position" which "had been made possible by a generous and much appreciated bequest made by a former member of the Faculty" (the late Professor Kennelly) for "applied mathematics, one of the fields in which Professor Kennelly himself had made notable scientific contributions." Kennelly is better known today for his work on the physics of the ionosphere than for any contributions to applied mathematics. His name is celebrated in the name of the Kennelly-Heaviside layer, one of the layers of ionized air that exist from 100 to 120 kilometers above the Earth's surface. (Also known as the E-layer, the Kennelly-Heaviside layer has the property of reflecting radio waves of certain frequencies back to the Earth.)

In a retrospective article on American mathematics during his lifetime, Marshall Stone—a distinguished member of the Mathematics Department at Harvard and later of the University of Chicago—recalled his role in making funds available to Aiken. Stone's own interest in harmonic analysis had prepared him to be sympathetic to Aiken. Stone, indeed, had included in a course on that subject "a discussion of integraphs, harmonic analyzers, and other analogue computers." During his work on "multiplicative systems needed for studying the axioms of Boolean algebra," Stone recalled, he had become aware that "it might be possible to carry out the time-consuming calculations more quickly and with less risk of error if an electrical device easily adapted to changes in the multiplication table could be built." Stone was, accordingly, an important ally when it was proposed that Aiken be supported with funds bequeathed to Harvard by Arthur Kennelly. According to the terms of the Kennelly bequest, any expenditure of funds had to be approved by the Harvard Mathematics Department. The department, Stone reported, "voted unanimously to support the grant" for Aiken and "thus lent its support to the development of the first modern digital computer."

Aiken's Harvard appointment by way of the Kennelly bequest was rather appropriate for a special reason: Aiken, while still a graduate student, had served for a short while as a Washington lobbyist, striving to preserve for research purposes the right of universities to bounce radio waves off the Kennelly-Heaviside layer. This research required that radio signals be sent and received automatically—that is, without a human operator.

Aiken's lobbying efforts are recorded in some recollections written down after Aiken's death by Harry Mimno, who had been the junior member of the Cruft group when Aiken was in graduate school and who later taught in Aiken's computer science program. During the years when Aiken was a Harvard graduate student, Mimno was engaged in research on the ionosphere. He sent up short-wave radio signals and measured the time it took for them to reach the Kennelly-Heaviside layer and be bounced back. In order to obtain a continuous record, it was necessary to send up signals at regular intervals during each 24-hour period. These signals were sent out automatically; it would have been far too costly to have to employ operators to be on hand all through the day and night. At that time, government regulations required that an operator be on hand whenever any radio signals were being sent out. Universities engaged in research were, however, allowed to send and receive signals without an operator. Mimno, concerned that Congress might take away the universities' special experimental status, sent Aiken to Washington to lobby on the universities' behalf.

Aiken was a good choice for this assignment. He was older than other students. Indeed, always neat and well dressed, he did not look like a student at all. He was tall and erect in carriage, and he spoke with a midwestern accent. Coming as he did from an industrial-commercial background, Aiken could speak the language of the marketplace as well as plead for the needs of academic science. Mimno was convinced that the Washington bureaucrats would be impressed by Aiken's "relative maturity, his vigorous personality, and his obvious assurance when convinced of the merit of a proposal."

When Aiken got to Washington, however, he found that the Federal Communications Commission's lawyers were opposed to his proposal; they simply did not believe that Congress had given them authority to make exceptions to the rule. Aiken made additional trips to Washington and tried another tack. He eventually succeeded in his mission, getting a Massachusetts congressman to introduce a bill that would guarantee the needed kind of operation for experimenters. As often happens in such cases, however, the bill "died on the calendar" at the end of the congressional session.

In the next session, Aiken went into action with typical vigor and efficiency. This time he carefully analyzed the factors that made for the successful passage of a bill such as the one that would grant Harvard and other research institutions an exception to the rule. When it became certain that, thanks to Aiken's astute maneuvering, the bill would pass, the FCC decided to use the opportunity to extend the terms of the enabling act so as to give it greater flexibility in dealing with automated signaling devices. As Mimno noted, Aiken's activities had the later beneficial result of making "all communications satellites viable."

# 4
## First Steps Toward a New Type of Calculating Machine

In 1973, when Henry Tropp and I were planning our interview with Aiken, we agreed that a primary topic to explore was how he had come to design the giant machine that launched him into the world of computers. Accordingly, soon after the interview began, I asked Aiken directly what had shifted him away from a career as an electron physicist and turned him into a computer scientist—what had started him in that new and apparently somewhat unrelated professional direction. Aiken replied without a moment's hesitation. The subject of his doctoral thesis, he explained, was space charge, "a field where one runs into differential equations in cylindrical coordinates (or in a parallel case, in rectangular coordinates)—in nonlinear terms, of course." Actually, he continued, "the object of the thesis almost became solving nonlinear [differential] equations: not completely, but there was some of that in it." The only methods then available for numerical solutions of problems like his made use of hand calculations, which were "extremely time consuming." It became apparent—"at once," according to Aiken—that the labor of calculating "could be mechanized and programmed"—that "an individual didn't have to do this."

In those days, individuals (usually women) performing calculations were generally known as "computers"; desk machines, whether mechanically operated or powered by electric motors, were called "calculators." Those who performed calculations made extensive use of numerical tables and tables of functions, including formulas for differentiating and integrating. Aiken's doctoral dissertation included a photographic reproduction of a portion of a table of Bessel functions reproduced from a widely used mathematical handbook compiled by Eugen Jahnke and Fritz Emde.

Most of the above-mentioned desk machines, mechanical or electromechanical, were essentially adders or accumulators. Those in widest use in the United States were made by Burroughs, Monroe, Marchant, and Frieden.[1] They multiplied by the method of successive additions with a shift of the carriage. Some people engaged in computation used the "Millionaire," a direct-multiplication machine housed in an oblong brass box. Most Millionaires were purely mechanical, but some had been converted to run on electricity. For certain types of scientific calculation there were various differential analyzers, devised by Vannevar Bush; these were essentially large-scale interconnected integrators or super-planimeters.

At this time, however, there was one major facility with some capacity for large-scale calculation: the computation center at Columbia University. Begun as the Columbia University Statistical Bureau, this facility was later known as Columbia's Watson Scientific Computing Laboratory. In the late 1930s this center was under the direction of Wallace Eckert, who was using IBM punched-card tabulating machines of various sorts, some modified so as to be used more efficiently in calculation. Eckert was a Ph.D. in astronomy from Yale, where he had worked under E. W. Brown, famous for his investigations of the motion of the moon. One prominent member of the staff who later had an important career in computers was Herbert Grosch, also a Ph.D. in astronomy. At the Columbia center, Eckert and his associates were busy computing problems of lunar orbits according to the equations developed by Brown.

In England, just before World War II, L. J. Comrie had established himself as an outstanding figure in the world of machine calculation. Comrie had an interesting career. A New Zealander by birth and upbringing, he was wounded in World War I, losing a leg as a result. As part of his rehabilitation, he was taught the art of computing by Karl Pearson, then one of the world's leading statisticians. One of Comrie's fortes was to adapt business machines to scientific calculation. Comrie was Superintendent of the British Nautical Almanac office for a number of years before founding Scientific Computing Services, Ltd.[2]

1. At Harvard we tended to use Monroes and Marchants; there were no European machines of the Odhner pinwheel type. We were aware of the existence of more complex machines, originally made for business, such as those of Brunsviga and Mercedes-Euklid.

2. For more on Comrie, see chapters 8, 16, and 19.

No machine (or combination of machines) of the 1930s was adequate to the task of solving the kinds of problems that arose in Aiken's thesis research in any reasonable time. Aiken's experience with this inadequacy, and his awareness of other problems in the sciences and in engineering that could not be readily solved by means of hand calculation and desk machines, stimulated him to invent a means of solving mathematical problems by machine. Not only did he foresee that such a machine would be useful for various branches of science and engineering; he even foresaw applications in the social sciences.

One of Aiken's fellow graduate students, Ivan Getting (later Raytheon's chief scientist), says in his memoirs that Aiken once devised an electromechanical calculator using old "telephone" relays. According to Getting, some time after Aiken paid his first visit to IBM he was given some relays and counters on the supposition that parts of the projected machine would be constructed at Harvard. This plan was rather quickly abandoned, however, and eventually IBM did all the construction in its Endicott plant.[3]

Ivan Getting's reference to "telephone relays" is notable. In the 1930s and the 1940s, the commonest use of relays was in the telephone system, whereas earlier the most important application of relays had been in long-distance telegraphy. Many types of electrical circuits, both those used in scientific experiments and in commercial applications, used the kind of relays that had been standardized by the telephone company. When I was a graduate student, we always referred to this class of relays as "telephone relays," even if the relays in question had not been made for, or ever used by, the telephone company. Thus, even if the relays Getting remembered had been manufactured by IBM, they would have been known informally as "telephone relays." Getting's recollections of "telephone relays" should not be interpreted as evidence that Aiken had made contact with the telephone company.

A relay is essentially an electromechanical switch, the primary components of which are an electromagnet, a pivoting iron bar or plate known as the armature, and a spring. The spring holds the switch open, the normal position of the armature. When an electric current

---

3. Ivan Getting does not give a date, and there is no way of determining either the date of this event or the circumstances. Getting's recollections probably concern this early phase of construction of Aiken's machine and should be dated to a time after IBM became interested in Aiken's proposed supercalculator. Getting was evidently not referring to a pre-IBM attempt by Aiken to build his own calculator.

energizes the coil of the electromagnet, the pivoting bar is pulled toward the electromagnet, closing the switch. When the current ceases to flow through the coil, the electromagnet loses its magnetic power and the spring pulls the bar back to the open position. Thus, flow and lack of flow of current through the coil of the relay correspond, respectively, to the on and off positions of a switch.

Relays came into prominent use with the introduction of the electromagnetic telegraph in the middle of the nineteenth century. A telegraph signal sent over long distances may produce a current too weak to activate the "sounder" at the receiving end, even though the current may be strong enough to operate a sensitive relay. This relay acts as an electrical switch, turning on or cutting off a local current of any desired strength which operates the sounder. Relays were used in a somewhat similar fashion in long-distance telephony to convert weak electrical signals into stronger ones. They were also used in the telephone industry in the operation of switchboards. Aiken's first two machines made extensive use of relays or electromagnetic switches, as did other super-calculators of the 1940s.

Another published reference to an early Aiken calculator appears in *The Analytical Engine,* a popular book on computers written by the physicist Jeremy Bernstein. Bernstein writes that Aiken had built (or was beginning to build) some kind of analog or relay calculator before he made contact with IBM in 1937. When I wrote to Bernstein asking for more information about this early Aiken calculator, he replied that he had learned of it directly from Aiken in the course of an interview. Unfortunately, Bernstein no longer has the notes from that interview.

In my research I have found no evidence that Aiken ever designed or built any kind of calculator before 1936, when the problems of his thesis set him on the path of calculator design that would eventually produce the ASCC/Mark I. Nor is there any hint that Aiken ever planned or constructed an analog machine before working up his proposal for a giant digital calculator. Perhaps Bernstein had in mind the hydraulic analogue of a problem in electron flow that made such a strong impression on both Jim Hooper and Ed Purcell. (On this device, see chapter 3.)

Aiken's steps toward the construction of a giant mechanical calculating device are traced in a memorandum, written shortly after Aiken's death, in which Harry Mimno recalled Aiken's achievements. Mimno describes a discussion he had with Aiken concerning a possible means of automating the analysis of the results of ionosphere experiments. Suddenly, Aiken was discussing "the possibility of directing the activities of scores of computing units, initially talking in terms of racks of interrelated machines, each roughly equivalent to the Marchant or Monroe desk computers of that day and age." At this time, according to Mimno, Aiken had already shifted his primary focus of research attention to "large-scale automated computers." Mimno's own role, he recorded, was "that of an enthusiastic and fascinated listener." The one aspect of Mimno's account that stands out is Aiken's proposal of "governing the program of the individual units" by means of paper tape. Mimno did not specify the date, but clearly Aiken already had in mind an automated system of computing that would be programmed or controlled to do a series of operations automatically in sequence—three of the fundamental principles of the computing machine he eventually proposed to IBM. Mimno further records that there was some discussion of whether "the program-tape" should be "sensed pneumatically" (in the manner of a roll of music fed into a player piano) or "sensed electrically" (in the manner of "the automatic-telephone machinery then current"). After "considerable discussion," according to Mimno, the "electric-sensitive approach won out."

Some time in 1936 or possibly in early 1937, Aiken's thinking about machine calculation had gone well beyond the immediate needs of his thesis. He was now concerned with general aspects of large-scale calculation by machine. At this point there are several problems in chronology that I have not been able to solve. When Tropp and I interviewed Aiken, I reserved a few questions about chronology for our proposed later sessions, which—alas—we were unable to conduct because Aiken died soon after our first meetings. However, when Tropp and I asked Aiken about early documents and letters, he told us that he no longer had any early documents. When he left Harvard in 1941 for active duty in the Navy, his papers were packed up in boxes, which were lost or thrown out during the war.

We cannot, therefore, be certain about either the dates or the details of the very first plans for the giant calculator Aiken conceived. Was it to be electromechanical, using relays, from the start? Had Aiken ever contemplated a gigantic mechanical device run by electricity? Did he ever conceive that his brainchild might use vacuum tubes?

One date is certain, however: by April 1937, at the very latest, Aiken had progressed sufficiently far in his general thinking and design to be ready to seek support from industry. In view of Aiken's work habits, it is not difficult to imagine that he drew up a careful memorandum stating the need for such a machine, its principal features, its mode of operation, and its general method of solving problems. His philosophy was later expressed in an assignment written for one of the classes at Harvard: the design of an inexpensive laboratory computer (or calculator). "The 'design'" of a "computing machine," the students were informed, "is understood to consist in the outlining of its general specifications and the carrying through of a rational determination of its functions, but does not include the actual engineering design of component units." In this clear statement, as was often the case for Aiken, the logic of the machine, the mathematical operations, and the general architecture had precedence over the technological specifications and the choice of components. That is, Aiken assumed that the design of a computing machine included only the specification of the logic or the sequence of controlled operations that the machine was to be programmed to perform. In contrast with the present concept of computer architecture, such a design is not concerned with the "physical or hardware structure of computer systems and the attributes of the various parts thereof" or with the consequent physical or hardware problem of "how these parts are connected." To judge from all the information available, Aiken's design would not have specified what particular components (or even what types of components—mechanical, electromechanical, electronic) would be used, or how the various components of the machine would be linked. He would have specified the need to perform certain types of mathematical operations and a means of programming them so that they would be performed in a certain predetermined sequence. He also would have indicated the need to store certain tables of numerical data. These specifications would have been definite, but they would not necessarily have been confined to any particular type of functioning elements. Thus, the design would apply equally to a machine constructed of mechanical, electromechanical, or electronic components, or any combination thereof.

Since there is no firm documentary evidence of when Aiken completed the general design of the machine that he wished to have built, we can only proceed by inference from a single fixed date: 22 April 1937, when Aiken formally presented his plans to the first prospective builder, the Monroe Calculating Company. Hence, Aiken's actual design had certainly reached an advanced stage by the end of April 1937, and very likely some time earlier. Aiken was not the sort of person to complete an assignment at the last minute. Thus, it is most likely that his general design was completed and ready to be proposed in a formal document some time during the early months of 1937.

# 5
## *An Unsuccessful Attempt to Get the Machine Built*

Once Aiken had completed the general design of his proposed machine, his next step was to find a company willing to build it. During the course of our interview, Aiken explained to Tropp and me that, because of the size and complexity of his proposed machine, only a large manufacturer of calculators or business machines could have been induced to produce it. Because in a sense Aiken's machine could be considered a gigantic calculator, he naturally chose to start his search for support with America's foremost manufacturer of calculators. His first choice was the Monroe Calculating Machine Company. Armed with his document of specifications, Aiken obtained an interview, which took place on 22 April 1937, with George C. Chase, a distinguished inventor in the calculator field who was then Monroe's director of research.

Chase had joined the staff of Monroe in 1917, 20 years before Aiken's visit. Under his direction, Monroe became a leader in the field of calculating machines, winning the John Price Wetherill Medal of the Franklin Institute in 1932. The citation took note that the special success of Monroe's mechanical desktop calculators in implementing "the four basic rules of arithmetic: addition, subtraction, multiplication, and division was attained through Mr. Chase's inventions."

Chase later published an account of Aiken's visit that includes the date, 22 April 1937, so we know that Aiken had his design worked out by that time. Chase's record of Aiken's visit is very detailed. Aiken outlined his conception of the machine and "explained what it could accomplish in the fields of mathematics, science, and sociology." Aiken told Chase that "certain branches of science had reached a barrier that could not be passed until means could be found to solve mathematical problems too large to be undertaken with the then-known computing equipment." Although Aiken referred to "the construction of an electro-mechanical machine," he did not specify what actual components were to be used. Chase was quite emphatic on this point. The "plan he outlined," Chase wrote, "was not restricted to any specific type of mechanism." Rather, Aiken's design "embraced a broad coordination of components that could be resolved by various constructive mediums." This accords well with Aiken's philosophy, embodied in the Harvard assignment quoted in chapter 4 above. It is true that Aiken did mention an "electro-mechanical machine"; however, this statement must be understood in terms of Monroe's product line. Because Monroe specialized in mechanical and electromechanical machines, Aiken would not have suggested vacuum tubes or other electronic elements as the primary operating units.

Aiken's attempt to elicit the support of Monroe came up rather early during the interview,[1] when I pressed him to explain why he had chosen to build Mark I out of electromechanical parts. After all, his thesis was on vacuum tubes, on space charge, and his own graduate specialty was the field of electronics. Why, I wanted to know, did he even consider electromechanical systems rather than electronic systems? Why had he not contemplated using vacuum tubes? I will confess that I expected Aiken to frame his reply in terms of his great and often-expressed ideal: reliability. I will even confess that I asked the question less as a means of obtaining information than as an opportunity to record on tape—direct from Aiken's mouth—his thundering condemnation of vacuum tubes as unreliable and his preference for slower and more reliable relays. It was only much later that, thanks primarily to a little tutorial given to me by Bob Campbell and to the insightful comments of Maurice Wilkes, I came to understand that Aiken's study of the physics of vacuum tubes was only indirectly related to the use of vacuum tubes in designing electronic circuits. In fact, in a statement written in 1945, Aiken reviewed the goals of education in the Cruft Laboratory, where the "plan for instruction" had been designed around "basic scientific material of communication engineering" and "much of the allied branches of science." There was

1. Chase's published account of Aiken's visit was the final entry in a booklet he wrote on the history of calculating methods and calculating machines. On 24 February 1973, when Henry Tropp and I conducted our interview, I happened to have a copy of Chase's booklet with me. Bernard Galler, the editor of a new journal, *Annals of the History of Computing*, had decided to reprint this work and had asked me to write an introduction. Thus, I was prepared to discuss that first attempt to find support for the project.

"no attempt to apply this material to specific engineering problems." Instead, the program had been devoted exclusively to "the elucidation of fundamental scientific principles." For the purposes of computation, however, what was needed was not the scientific principles underlying circuitry, not a knowledge of the physics of space charge, but rather some experience in the design of high-speed pulse circuitry—an area in which Aiken had little or no experience.

In a talk given in Sweden and in Germany in 1956, Aiken recalled that, in his senior year at the University of Wisconsin, "there was offered for the first time a course called 'thermionic vacuum tubes.'" Aiken didn't explore this new field, however, because his professors advised him that he "would do far better" if he "took the course in transformer design rather than this new and untried subject."

Aiken had extensive training and experience in electrical engineering, but none in the design and technology of the kind of circuitry needed for a large-scale computing machine. Indeed, as he made clear during the 1973 interview, he had never been wedded to any particular technology. To my complete astonishment, he frankly declared that one technology would have served as well as another, so long as it enabled a large-scale programmed calculator to perform the desired sequence of mathematical operations.

Aiken told Tropp and me that he had been fully aware that making his computer a reality would require "money and a lot of it." Since he was not then—nor was he ever—primarily interested in technological innovation, it had seemed to him that the most sensible course was to "build the first machine out of somebody's existing parts," rather than to have to invent or construct parts on his own. Electromechanical relays and step switches were already in wide use, teletype had been developed, and there was punched tape or punched cards for input. "The tape," he said, "was harder to edit and you couldn't sort with it, but nevertheless it would work and it had advantages." These "different techniques—printing telegraph techniques, telephone switching techniques, communications industry techniques—were all grist for my mill." At that time, Aiken said, he was "largely a promoter, trying to find out where to get these pieces so that the machine could be put together." Aiken then told us that his "first step" was to go to "the Monroe Calculating Machine Company." Aiken could not at once recollect "the name of the charming man I met there." "Mr. Chase," I told him—"George Chase." When Aiken expressed astonishment that I knew Chase's name, I told him that I was preparing a reprint of his anecdotal history of computing, and that on reading it I had discovered his account of Aiken's visit. I showed Aiken a photocopy of Chase's booklet, and I read aloud the sentences I have just quoted about Aiken's plans. "He's just saying what I said a moment ago, only much better," Aiken commented, adding that he "went to Chase and . . . did just what he said."

Aiken's plan, according to Chase,

provided automatic computation in:—the four rules of arithmetic; preestablished sequence control; storage and memory of installed or computed values; sequence control that could automatically respond to computed results or symbols, together with a printed record of all that transpires within the machine; and a recording of all the computed results.

These are some of the main features that were embodied in the eventual machine that IBM's engineers designed and built to meet Aiken's specifications. Indeed, this passage reads as if taken from a document that Aiken either brought to the meeting and left with Chase or mailed to Chase afterwards. (No such document appears to have survived.)

In the course of the 1973 interview, Aiken said that Chase—"Chief Engineer at Monroe" and "a very, very, scholarly gentleman"—"took an almost immediate interest, and we kept up an association for quite a few years thereafter. He wanted, in the worst way, to build Mark I. He would supply me with the parts and we would collaborate and do it together, that's what he wanted to do." "He also foresaw what I did not," Aiken said. "I did not foresee the application to accounting as coming out of it, and he did."

Aiken was encouraged by Chase's enthusiastic support of his project. Chase, he told Tropp and me, "went to his management at Monroe and he did everything within his power to convince them that they should go ahead with this machine because, although it would be an expensive development." Chase had the vision and foresight to recognize that the proposed machine "would be invaluable in the company's business in later years." But, although "Chase could see this," his "management . . . after some months of discussion turned him down completely."

I was pleased to learn these details of Aiken's first attempt to get his machine built, but I was not completely satisfied. I wanted him to discuss the advantages and disadvantages of mechanical systems, electromagnetic devices, and vacuum-tube circuits. A little later in the interview, I returned to the subject of why Aiken had chosen to have his machine built of electromechanical components—why he had not made use of vacuum tubes. This time I stressed the fact that the choice of relays had always seemed astonishing to me in view of the fact that at Harvard Aiken had been a student of E. L. Chaffee, a specialist in vacuum tubes and vacuum-tube circuits. To be specific, I asked whether at one time he hadn't considered including quenching circuits using vacuum tubes in the first machine. Aiken replied:

Yes. But your question really is: since I had grown up in "space charge" in a laboratory like Cruft, why wasn't Mark I an electronic device? Again the answer is money. It was going to take a lot of money. Thousands and thousands of parts! It was clear that this thing could be done with electronic parts, too, using the techniques of the digital counters that had been made with vacuum tubes, just a few years before I started, for counting cosmic rays.

He concluded with this dramatic assertion:

But what it comes down to is this: if Monroe had decided to pay the bill, this thing would have been made out of mechanical parts. If RCA had been interested, it might have been electronic. And it was made out of tabulating machine parts because IBM was willing to pay the bill.

In many ways, this was the most fruitful revelation made in the course of the interview. It shows that Aiken was not at that time wedded to any particular technology, and that his top priority was not the choice of relays. It seems astonishing that the choice of the kind of machine to be built was determined solely by financial considerations, by the willingness of one or another company to put up money for the machine. This disdain for the technological components was, I believe, a very significant part of Aiken's intellectual makeup. This aspect of Aiken's system of values was a major factor in producing the eventual rift between him and IBM. Aiken never appreciated the degree to which the technology of IBM's product line may have made IBM the only company that would have undertaken to build Aiken's machine at that time. (When Eckert and Mauchly designed the ENIAC, they did not base it on any company's off-the-shelf technology; rather, they developed new types of circuitry and design for the special purpose they had in mind.)

Aiken's account in the interview would suggest that he made a kind of block diagram, being concerned primarily with the general features of the machine's architecture, the mathematical operations to be performed, and the programming of their sequence, and not caring too much about the specific nature of the individual components. Aiken, of course, would have been aware that many parts of the electromechanical functioning system could, if necessary, be replaced by mechanical components or by vacuum-tube circuits. Relays and diode vacuum tubes both act as on-off switches, vacuum tubes having the advantage of being many times faster in their operation. I interpret Aiken's statement to us as implying only that his primary goal was to get his dream machine built and that this overrode any considerations of technological detail.

If Monroe had decided to build Aiken's machine, it might well have used relays in various systems of mechanical linkages. It seems obvious in retrospect that this would have been a source of many problems and would have made for a machine less reliable than the one built by IBM. We know about the nature of such problems because they arose, as Robert Campbell has pointed out to me, in the operation of some early Remington Rand machines that used such mechanical devices as "Bowden wires" to "read" punched cards and pass on information. Again in retrospect, it seems beyond the bounds of any reasonable possibility that RCA would have been willing to build Aiken's machine at that time, since they had no technology ready for such application. In this sense Aiken needed IBM, whose technology included the use of punched cards, the accumulation of numerical data, and the transfer of numerical data from one register to another.

Another possible builder of the machine might have been the Bell Telephone Laboratories, which had much of the needed technology in hand and might possibly have developed a machine to conform to Aiken's design. We know that the Bell Telephone Laboratories later constructed a line of relay calculators inaugurated by George Stibitz. But the record shows that the Bell Telephone Laboratories were not at first interested in exploiting Stibitz's ideas, and that they began to produce advanced calculators only under the pressure of wartime needs.

For Aiken's dream to become reality it was fortunate that IBM was both technologically prepared to build the machine and willing to commit its resources to the assignment. It should be kept in mind that Aiken's machine required an engineering design effort of a kind and on a scale that IBM had never before attempted. It was not merely a matter of assembling ready-made components according to an established technology.

# 6
## Seeking Support from IBM

Monroe's decision not to support Aiken's project was certainly a blow, but Aiken must have been heartened by George Chase's enthusiasm for the new machine. Furthermore, it was Chase who suggested that Aiken turn to IBM. In the 1973 interview, Aiken told Tropp and me that when he had asked Chase "Whom should I see at IBM?" Chase had told him: "Why don't you see Professor Brown? He's at the Business School at Harvard. He's right there."

An applied mathematician and a trained astronomer with a Ph.D. from Yale in celestial mechanics, Theodore (Ted) Brown had long been interested in various applications of business machines and their use in solving scientific problems, and his graduate training had introduced him to aspects of mechanical calculation in science. He was, furthermore, a close associate of Thomas J. Watson.[1] According to Aiken, Brown "came to life" and sent him to see Harlow Shapley, Director of the Harvard College Observatory. Shapley, in turn, "came to life" and made the successful contact with IBM.

There is no reason to doubt that Chase suggested to Aiken that he make contact with IBM or that he directed the line of contact that eventually led Aiken to Brown. But the link was not Brown-to-Shapley-to-Watson, as Aiken remembered. Rather, the road to IBM began with Shapley. I found confirmation of this direct set of links in Shapley's correspondence, notably a letter written by Shapley to Watson after the contretemps at the dedication of the ASCC/Mark I. I also discovered, during the course of an interview with Brown, that he had indeed discussed Aiken's proposed machine with someone at IBM. But

---

1. Here and henceforth, "Thomas J. Watson" should be presumed to refer to the senior Watson. His son is always identified as Thomas J. Watson Jr.

the first direct discussion was not with Watson; it was with James Wares Bryce, IBM's chief engineer.

In a historical memorandum about Aiken and IBM, written almost immediately after the dedication, Francis E. Hamilton, the IBM engineer immediately in charge of the construction of the machine, recorded that Brown had spoken to Bryce about Aiken and his project before Aiken sent Bryce a formal request for a meeting. In the course of an interview, Hamilton recalled that Bryce—"known as the Father Engineer at that time"—was a "very bright man" with "several hundred patents" to his credit, but nonetheless "a very democratic sort of person." Hamilton also recorded that Bryce had an assistant named Halsey Dickinson; he remembered that both Bryce and Dickinson viewed Aiken's proposal very favorably. Hamilton also noted that at that time Clair Lake (the official supervisor of the design and construction of the new machine) "was a very good friend of Bryce's," and that the two of them discussed the calculator and "decided that Lake could handle the project."

Harlow Shapley, famous for his pioneering studies of the size and nature of our galaxy, was appointed director of the Harvard Observatory in 1921 at the age of 36 and quickly became a powerful figure in the scientific world. He took special delight in sponsoring projects by younger scientists and scholars, and he was just the right kind of person to advance a project like Aiken's. And Aiken's proposal must have had a special appeal for Shapley, who had long been interested in the use of machines in astronomical calculations. Shapley and Ted Brown knew each other because Brown had remained interested in astronomical problems, especially the use of machines in orbital calculations.

Brown himself was an interesting figure. When Owen Gingerich (Harvard's historian of astronomy) and I interviewed him, shortly before he died, he told us about his training at Yale as a mathematician and astronomer. He had done his graduate work at Yale on mathematical problems of celestial mechanics, working on aspects of the moon's motion under the direction of Ernest William Brown. E. W. Brown was celebrated for his studies of the complex motions of the moon. Two major computing enterprises were then underway to perform the calculations deriving from E. W. Brown's equations, one headed by L. J. Comrie in London and one by Wallace Eckert at Columbia University. Theodore Brown told us that he had first met Shapley when the latter spoke to a group of astronomers and graduate students at Yale on Eckert's use of modified IBM tabulating machinery to solve astronomical problems.

After getting his Ph.D., Theodore Brown came to Harvard in the hope of working with George Birkhoff, America's foremost mathematician, who was extending the methods Henri Poincaré had developed for treating problems in celestial mechanics by geometric or topological methods. The only Harvard post Brown could obtain, however, was in the Graduate School of Business Administration, where he eventually became Professor of Statistics, a position he held until his retirement. Part of his training as an astronomer had been in applications of statistics, and the business school apparently offered Brown a better source of income than either the math or the astronomy department. A longtime consultant to IBM, Brown became a great favorite of Thomas J. Watson.[2] Brown regularly conducted special training courses at IBM and often lectured to the IBM staff. He was a favorite speaker at IBM functions, such as the regular "corporate recognition" events. During our interview, Brown proudly showed Gingerich and me a gold watch and an original Hollerith counting wheel (mounted in a plush-lined box), both testimonial gifts from Watson. At the time when Aiken made contact with Brown, Brown was a member of the Advisory Board of the computation laboratory that Watson and IBM had set up at Columbia University and so was directly involved in the problems of scientific calculation by machine.

Brown's expertise and his long-term relationship with Watson and IBM made him the ideal person to introduce Aiken and his dream to IBM. Watson was always respectful of Ivy League academia to an extreme degree. Throughout his life he was noted for his genuine concern to use IBM's resources for educational purposes and especially for the advancement of science. I do not know at what stage Watson himself entered the picture. Possibly Brown mentioned the Aiken matter directly to Watson. But there is no doubt that Brown established the contact between Aiken and IBM's chief engineer, James Bryce, and that this was the inaugural step toward the construction of the Automatic Sequence Controlled Calculator.

Watson, a titan in his sphere, had just as forceful a personality as Aiken. When two such strong personalities as Howard Aiken and

2. Brown told Gingerich and me that Watson even offered to build him a house in Endicott, where Watson hoped to establish a campus much like a British college in appearance, if he would leave Harvard and join IBM full time.

Thomas Watson met, a terrible clash was inevitable. Thomas J. Watson Jr., who eventually succeeded his father as head of IBM, is quoted as saying "If my father and Howard Aiken had had revolvers, they would both have been dead."

Thomas John Watson was born in 1874 in East Campbell, near Corning in rural New York. After a year of studying business and accounting, Watson became a salesman of pianos and organs. He quickly learned that a successful salesman must dress the part, and later he enforced a rigid dress code on himself and his employees. Later on, when he had become head of IBM, he insisted that even the repairmen wear business suits, white shirts, and neckties. Traveling around selling pianos and organs, Watson learned that many salesmen—lonely and rootless—found solace in drink. Throughout the rest of his life, he worked to "divorce drinking from business," as his biographers put it, even to the point of becoming "accused of being too evangelical in the matter."

Eventually, Watson became a salesman for the National Cash Register Company. After his impressive sales record brought him to the attention of NCR's president, John H. Patterson, Patterson became in many ways the dominant figure in Watson's life and self-image. Patterson is sometimes considered the father of modern salesmanship. He taught his sales force that success requires motivation (which Patterson provided in the form of cash bonuses, raises, diamond stickpins, and gold-headed canes), education, and technique. He sent his salesmen to his own tailor, and he established company schools to teach employees the technical aspects of the product line and the art of selling. His goal, shared by Watson, was to "exalt the salesman" and to make "a business man out of him."

By 1910, Watson had become Patterson's right-hand man. Patterson—moody, jealous, and given to impulsive decisions—evidently resented Watson's ability and success and one day abruptly fired him. By this time, Watson, at age 40, had achieved enough of a reputation to be offered attractive posts by Frigidaire, Montgomery Ward, Remington Arms, and a Boston company that made water coolers. In the end, he chose the small and relatively unknown Computing-Tabulating-Recording Company.

The Computing-Tabulating-Recording (CTR) Company had recently been formed by the amalgamation of three small firms: the Computing Scale Company (which made scales and cheese slicers), the International Time Recording Company (which made clocks for keeping track of employees' hours on the job), and the Tabulating Machines Company (founded by Herman Hollerith to produce and develop punched-card tabulating machines of the kind originally invented to tabulate the US Census of 1890).

Watson found the tabulating business to be the most interesting part of the new conglomerate, although the other two branches made products that more closely resembled cash registers. After starting an experimental department dedicated to improving the tabulators, he sought to find new types of business enterprises where they might be used. By 1916 Watson had assembled a small staff of inventors that included the above-mentioned Clair Lake, later the engineer in charge of the design and construction of the Aiken machine.

Watson stressed loyalty. "Joining a company," he told his men, "is an act that calls for absolute loyalty in big matters and little ones." He preached that the company was the employee's "friend," and that a "family spirit" ("combined with vision and faith") was responsible, "perhaps more than anything else," for a company's success. In 1924, in recognition of the fact that business tabulating machines had become its main products, Watson gave his enterprise a new and grandiose name: International Business Machines Corporation.

Following the precepts of Patterson, Watson insisted on certain standards of conduct for all IBM employees. There was a strict policy of never mixing drinking with business, and every employee was expected to "dress the part." There were regular classes and organized conventions. The salesmen who had fulfilled their sales quotas—the Hundred Percent Club—were treated to great celebrations in major cities, where the company put them up in posh hotels such as New York's Waldorf Astoria.

According to Watson's authorized biographers, meetings of the sales force became "part camp meeting, circus, and Chatauqua, superimposed on the versatile form set by the NCR, all of which Watson adopted—the name, the bands and blare, the flags and group pictures, the daily newspaper, banquets at horseshoe tables, even the spirit of evangelism." There were company songs ("To Workers in Our Factories," "T-H-I-N-K That's Think") and even a company symphony (especially composed by Vittorio Giannini).

Watson, his biographers relate, "believed that engineering, like salesmanship, depended not only on laws but on will." Accordingly, for him "the first principle of science, as well as the first principle in the world of men, was enthusiasm." He would arbitrarily tell his engineers "Build it!" And "when they did, the machine often seemed to be a triumph of Dale Carnegie over Newton."

Everywhere throughout the business office were big signs reading THINK. Small placards with this word emblazoned on them were on every desk. At one time, the lawn had this word set forth in red tulips. It became the name of IBM's company magazine.

Watson not only advanced the design and the increased use of business machines; he also pioneered the art of salesmanship, building on what he had learned during his apprenticeship to Patterson. Watson taught his salesmen to sell results and not machines—to emphasize "applications, not hardware, the *why*, not the *how*." By the time Howard Aiken came to IBM with a proposal to build a new kind of calculating machine, IBM had grown from a small company to an entrepreneurial giant with a gross annual income of about $40 million. Watson was earning the highest corporate salary reported by the Bureau of Internal Revenue.

When Aiken put forward his proposal, James Bryce was IBM's senior engineer. Bryce, the only member of the staff of "inventors" in whom Watson had complete confidence, had joined CTR as a 20-year-old draftsman and engineer in 1900, the same year George Chase became associated with Monroe. He had begun his service as "supervisory engineer" of the time-recording plant in Endicott. Bryce was the holder of more than 400 patents. In 1936, on the occasion of the centenary of the US Patent Office, he was honored as one of the ten greatest living inventors.

James Bryce and Ted Brown had a close friendship, not merely an association through IBM. Of all the engineers and executives Aiken met at IBM, Bryce was the only one he admired without reservation.[3] Indeed, he was the only person at IBM of whom Aiken spoke without rancor.

Bryce had attended the City College of New York for 3 years, and for a long time was one of the few college-trained men in the technical divisions of IBM. In the course of the 1973 interview, Aiken told Tropp and me that Bryce was "a very astute inventor" and "a valuable senior advisor." If "you started down a road that didn't look very practical," Aiken went on, "he could put his finger on it just about that quick."

3. During the 1973 interview, Aiken showed Hank Tropp and me a copy of Charles Babbage's autobiography that Bryce had given him and read aloud the inscription that Bryce had written in this book.

There is no doubt that Aiken's high regard for Bryce derived in part from the very fact that Bryce approved of his project and recommended that the dream machine be built. Additionally, Aiken did not have to deal with Bryce on the nitty-gritty level, where day-to-day friction is apt to sour the best of relationships. Bryce's support of Aiken was certainly crucial to IBM's decision to build the new machine, though the ultimate decision would have rested with Watson.

As IBM's historian, Emerson Pugh, has noted, Watson's support for the Harvard calculator project had a practical facet that is not usually recognized. One of Watson's goals was "to establish with Harvard University a mutually beneficial relationship of the type the company had long enjoyed with Columbia." It was to "foster this relationship" that Bryce would endow the Harvard machine with "the company's most innovative arithmetic and storage devices." It was envisaged that IBM would continually upgrade the Harvard machine, "thus providing a test bed for new devices and computational methods as well as a facility in which customer needs in scientific computation could be assessed." It was even imagined that the Harvard collaboration would gain IBM the services of a professor skilled in the new science of electronics—perhaps a member of the Cruft group.

# 7
## *The Proposal for an Automatic Calculating Machine*

Soon after Aiken was introduced to James Wares Bryce (IBM's senior engineer), he prepared a formal proposal titled Proposed Automatic Calculating Machine. This landmark text was printed in somewhat altered form in *IEEE Spectrum* (August 1964: 62–66) and in Brian Randell's book *The Origins of Digital Computers*. In *Makin' Numbers* it is presented as Aiken wrote it.

The Harvard University Archives contain at least three copies produced by some mimeograph-like machine. One copy of the proposal is in the files of Harvard's School of Engineering. Another is among the university's presidential papers for 1938, in a file marked Physics. Yet a third is among the Aiken papers.

The copy in the files of the School of Engineering is signed "Howard H. Aiken" and is dated, in Aiken's hand, "January 17, 1938"—apparently the day that Aiken officially delivered a copy to Dean Harald Westergaard. The copy in the Aiken files is dated, in pencil, "November 1937," about six months after Aiken's visit with George Chase at Monroe. The earlier date is confirmed by the reference to the proposal at the end of chapter 2 of the Manual of Operation for the machine. It is also supported by the fact that Aiken made his first contact with IBM in early November of 1937, when, after an exchange of letters, he presented his ideas to Bryce. It seems beyond the bounds of possibility that Aiken would have made contact with Bryce without a formal proposal in hand.

There is no way of telling whether the document in the archives is the one prepared for Chase or whether it is a later version, updated and revised for Bryce. However, the presence of at least three specific references to IBM components or IBM technology suggests that Aiken prepared this version after Chase recommended that he turn to IBM. In that event, the earliest possible date for the composition of this proposal is the summer of 1937, after Monroe turned down the project but well before Aiken saw Bryce at IBM. It may also be noted that Aiken does not refer to "sociology," the term used by Chase, but rather to "economics."

The version presented in *IEEE Spectrum* and in Randell's anthology differs in some respects from the document in the archives. For example, where the original refers to "International Business Machines," the reprinted version was changed (quite oddly) to "IBM machines." In the section titled "Present Conceptions of the Apparatus," there were numerous editorial changes—for example, Aiken's displayed and numbered lists lost their numbers and were converted into paragraphs.

Aiken's original proposal occupies 23 double-spaced typed pages. It opens with a brief history of "aids to calculation," a discussion of Babbage's engines, and a brief account of Hollerith's invention of punched-card "tabulating, counting, sorting, and arithmetical machinery." Aiken observes that the machines "manufactured by the International Business Machines Company" have made it possible to do "daily in the accounting offices of industrial enterprises all over the world"—the "things Babbage wished to accomplish."

Aiken then turns to the "need for more powerful calculating methods in the mathematical and physical sciences." He refers to the "new and useful functions"—most of them "defined by infinite series or other infinite processes" and largely "inadequately tabulated," with the result that "their application to scientific problems is retarded." Another need, he notes, is related to the fact that many "recent scientific developments" ("including such devices as the thermionic vacuum tube") are based on "nonlinear effects." The differential equations arising in the study of such phenomena "defy all methods available for their integration" other than "expansions in infinite series and numerical integration" involving "enormous amounts of computational labor." Here Aiken obviously had in mind experiences he had had while doing the research for his Ph.D. dissertation.

Aiken refers to the "present development of theoretical physics" (notably wave mechanics) as indicating that "the future of the physical sciences rests in mathematical reasoning directed by experiment." There are problems beyond our capacity to solve, he writes, "not because of theoretical difficulty, but because of insufficient means of mechanical computation." He gives as a specific example "the study of the ionosphere," which requires numerical results. The mathematical expressions, he says, are too long and too complicated to be solved by conventional methods or by other methods then available.

At this point Aiken introduces a really practical application: a study of the physics of the upper atmosphere, based on ionosphere research. On this, he says, "rests the future of radio communication and television." Here Aiken was drawing on his earlier association with Harry Mimno and on his own lobbying activities in Washington. Some other areas in which he suggests the proposed machine might be used are "astronomy, the theory of relativity, and even the rapidly growing science of mathematical economy." The reference to mathematical economics makes more sense than George Chase's statement that Aiken envisaged that his machine might by useful for solving problems in "sociology."

In the next section of his proposal Aiken specified four design features that differentiated ordinary punched-card accounting "machinery" and "calculating machinery as required in the sciences":

- A machine intended for mathematics must be "able to handle both positive and negative quantities," whereas "accounting machinery" is designed "almost entirely" for "problems of positive numbers."
- Calculating machinery for mathematical purposes must be "able to supply and utilize" many kinds of transcendental functions (e.g., trigonometric functions), elliptic functions, Bessel functions, and probability functions.
- For mathematics, a calculating machine should be "fully automatic in its operation once a process is established." In calculating the value of a function in its expansion in a series, the evaluation of a formula, or numerical integration (in solution of a differential equation), the process, once established, must continue "indefinitely until the range of the independent variables is covered"—usually "by successive equal steps."
- Calculating machinery designed for mathematics should be "capable of computing lines instead of columns," since often, in the numerical solution of a differential equation, the computation of a value will be found to depend on preceding values. This is actually "the reverse" of the way in which "existing calculating machinery" is capable of evaluating a function by steps.

Aiken concludes this section with the bold statement that these four features are "all that are required" to convert existing punched-card calculating machines ("such as those manufactured by the International Business Machines Company") into machines "specially adapted for scientific purposes." Aiken concedes, however, that "because of the greater complexity of scientific problems as compared to accounting problems, the numerical arithmetical elements involved would have to be greatly increased." We cannot tell whether Aiken's simple expression of optimism is to be taken as an index of his technical innocence. Perhaps he was making the practical problem of design and construction seem simple in order to convince the executives of IBM that his project was feasible. On the other hand, he was evidently convinced that his machine would require no more than a large-scale conversion and assemblage of existing commercial elements and not the design and production of wholly new and different working parts, producing a machine that was actually new in its basic conceptions and modes of operation. In the event, this distinction was to be a central issue in the strong divergence of opinion as to whether Aiken or the IBM engineers should receive primary credit for the invention.

Aiken's proposal specifies sixteen mathematical operations that should be built into the machine, including the four fundamental operations of arithmetic, parentheses and brackets, integral and fractional powers of numbers, logarithms and antilogarithms (base 10 and other bases), trigonometric and anti-trigonometric and hyperbolic and anti-hyperbolic functions, and superior transcendentals. These, he observes, will empower the machine to evaluate formulas and tabulate the results, to compute series, to solve ordinary differential equations of the first and second orders, and to perform both numerical integration and numerical differentiation.

Just as in his presentation to Chase, Aiken does not specify particular machine components that might be used in these operations. Rather, he mentions a set of "mathematical processes" that could be "the basis of design of an automatic calculating machine." Among the sample problems he introduces is finding the cube root of 5 by means of a method he had worked up in an equation for finding the $r$th root of $Q$ by iteration of the expression

$$x_n = \left(1 - \frac{1}{r}\right) x_{n-1} + \frac{Q}{r x_{n-1}^{r-1}}.$$

In another problem, Aiken shows how to evaluate $\sin(a + h)$ by "MacLauren's" theorem,[1] expanding into the series

---

1. Aiken's misspelling of Colin Maclaurin's name was not corrected by whoever made the other alterations in the reprinted version.

$$\sin(a+h) = \sin a + \frac{\cos a}{1}h - \frac{\sin a}{2}h^2 - \frac{\cos a}{6}h^3 + \frac{\sin a}{24}h^4 - \ldots$$

and computing "at most 10 terms." He shows the steps that lead to computation for various closed and infinite series, and he gives an example of how the roots of algebraic and transcendental equations could be found by successive mechanical steps. In these presentations Aiken refers, without explanation, to "methods of numerical solution . . . , such as those of Adams, Runge-Kutta, and others," and to "the rules of Simpson, Weddle, Gauss, and others."

Turning to "mechanical operations," Aiken notes that all the problems he has just discussed lead to numerical solutions which the machine could produce by "a repetitive process involving the fundamental rules of arithmetic." He then lists the major operations that current IBM calculative machines can perform, including

$A + B = F$,

$A - B = F$,

$AB + C = F$,

$AB + C + D = F$,

$A + B + C = F$,

$A - B - C = F$,

and

$A + B - C = F$,

where $A$, $B$, $C$, and $D$ are "tabulations of numbers on punched cards" and where $F$ is the result, also "on punched cards." $F$ cards can either be printed out or used as inputs (a new $A$, $B$, $C$, or $D$) in a further calculation. Any of these specified sets of calculations are to be determined "by means of electrical wiring on a plug board." Aiken once again declares that "all the operations described" could be "accomplished by . . . existing [IBM] machines when equipped with suitable controls." He concludes that "the whole problem of design of an automatic calculating machine suitable for mathematical operations is thus reduced to a problem of suitable control design, and even this problem has been solved for simple arithmetical operations." He then stresses two additional features of the proposed control mechanism: machine switching and "continuous perforated tape" in place of the traditional punched cards.

In the conclusion of his proposal, Aiken estimates that the machine will multiply a three-significant-figure number by an eight-significant-figure number (zeros not counted) in about 3 seconds. Multiplying an eight-significant-figure number by another eight-significant-figure number will take about 5 seconds. He proposes that there be 23 number positions (10 to the left of the decimal point and 12 to the right, and an extra position to indicate plus or minus). Clearly he was thinking of a fixed rather than a floating decimal point. Aiken is reported to have said later that his reason for having 23 digits was that he intended to recompute the planetary orbits.

Aiken proposes that the results of computations be printed out in tabular form so that they can then be printed by photolithography, thus eliminating errors that arise when numbers are copied by hand from a machine and those that occur during the various stages of typing, typesetting, proofreading, and printing from type.[2]

"No further effort will be made here to describe the mechanism of the International Business Machines," Aiken writes near the end of section VI. "Suffice it to say that all the operations described in the last section can be accomplished by these existing machines when equipped with suitable controls, and assembled in sufficient number. The whole problem of design of an automatic calculating machine suitable for mathematical operations is thus reduced to a problem of suitable control design, and even this problem has been solved for simple arithmetical operations."

The requirements for functional elements are very general and do not refer uniquely to any specific IBM devices. Aiken merely required that the functional elements be digital; that they be capable of performing the four fundamental operations of arithmetic; and that they be linked and controlled so as to perform their operations in a predetermined sequence, to store numbers (either constants or intermediate results) and introduce them at a specified stage in the automatic

2. A century earlier, Charles Babbage had devised a procedure, based on a different technology, that also would serve to eliminate these sources of human error. Indeed, this part of Babbage's schemes may have been one of his most original and important ideas in the art and technique of computing. The way in which Aiken's eventual machine produced its output is described in chapter 16 below.

sequence, and then to print out the final results in tabular form. IBM engineers could readily understand the function of each of the operative elements and could design systems in which these operations would be performed in sequence according to predetermined commands entered on punched cards or perforated tape. These engineers transformed Aiken's ideas and specifications into a set of functional elements that provided a basis for designing and constructing an actual machine.

Today, as the *Encyclopedia of Computer Science and Engineering* says, we use the term "computer architecture" to designate "the physical or hardware structure of computer systems and the attributes of the various parts thereof," and the way in which "these parts are interconnected." Computer architecture has developed from "computer design," which in its early years was seen as a branch of electrical or electronic engineering, dealing "primarily with electrical and electronic circuits and their organization into a computer system." The newer term, "computer architecture," implies an "amalgam of both electronic and software elements" and the philosophy that a successful design requires an "integration of both of these elements." From this point of view, what Aiken provided was perhaps more like a preliminary set of software requirements associated with a kind of "block diagram." This differs from today's concept of computer architecture, which is seen as "an element of the interdisciplinary field of computer science," along with such topics as "operations research, electrical and electronic engineering, and even solid-state physics." Aiken's presentation left the actual details of the architecture to be designed by IBM engineers working closely with Aiken.

As I have noted, in the 1930s the word "computer" did not signify a machine but rather a person doing computations with a desktop mechanical or electromechanical machine of a type generally called a "calculator." Aiken, in his proposal, spoke of an "automatic calculator," which he "visualized" as "a switch-board on which are mounted various pieces of calculating machine apparatus." Soon thereafter, the engineers at IBM and the Harvard authorities began to refer to the machine as a "calculating plant." That, in fact, was the name IBM used in its contract with Harvard.

# 8
## Aiken's Background in Computing and Knowledge of Babbage's Machines

One of the persistent beliefs about Aiken is that his thoughts on computing and computers, and the design of the ASCC/Mark I, were strongly influenced by the concepts and proposals of Charles Babbage. A typical expression of this belief appeared in the British scientific journal *Nature* in a 1946 review of the Manual of Operation for the above-mentioned machine. This review, written by L. J. Comrie, appeared under the title "Babbage's Dream, Come True."[1] Aiken would not have been displeased; he liked to consider himself Babbage's heir, and he often pointed out links between his ideas and machines and Babbage's. It would have been obvious to any reader of the Manual of Operation that Aiken considered Babbage his intellectual "father." There are quotations from Babbage, a fulsome description of Babbage's contribution to the computer, and even a portrait of Babbage. In *Faster Than Thought*, an influential volume on the state of the computer first published in 1953,[2] the chapter on "Computers in America" describes the "Harvard Mark I Calculator" as "the first machine actually to be built which exploits the principles of the

---

1. Aiken had, more or less, invited others to make such a connection by praising Babbage in the "Historical Introduction," which actually concluded with the sentiment that "though his [Babbage's] principles were theoretically sound and though he was successful to a limited extent, it remained for the twentieth century and the evolution of advanced mechanical and electrical engineering to bring his ideas into being."

2. *Faster Than Thought* had as its frontispiece a portrait of "Ada Augusta, The Countess of Lovelace," and its introductory chapter stressed Charles Babbage's great pioneering efforts. Also important in the Babbage revival was *Charles Babbage and His Calculating Engines*, ed. P. and E. Morrison (Dover, 1961).

analytical engine as they were conceived by Babbage a hundred years before."

Aiken's biographer and historians of the computer are faced with the problems of determining when Aiken first encountered Babbage's ideas and evaluating the influence of Babbage's writings on Aiken's thinking. Exploration of these related topics has brought out some real surprises and has revealed the stages of Aiken's education in the use of computing devices.

In evaluating Babbage's possible influence on Aiken, one must keep in mind the enormous gulf between today's image of Babbage as a computer pioneer and his relative obscurity before the 1940s. Today every historian of technology, every computer scientist, and every computer buff is aware of the prescient nature of Babbage's ideas and of the machines he either proposed or constructed in full or in part. The primary repository of documents on the history of the computer is called the Charles Babbage Institute. Recent years have witnessed the publication of a masterly eight-volume edition of Babbage's collected works, ably edited by Martin Campbell-Kelly, and there have been two separate volumes of selections from his writings. But in the 1930s Babbage was a rather obscure figure whose ideas were generally unknown to working engineers and applied mathematicians. His writings had been out of print for many decades and were to be found only in large research libraries. In my own student days, in the late 1930s, I knew of Charles Babbage, but not as the inventor of the difference engine or the analytical engine. Rather, Babbage had been a major figure in the Analytical Society, a small but influential propagandist organization that favored the Continental algorithm of the differential calculus over the British algorithm.[3] For many mathematicians and historians, not until after the advent of the computer was Babbage well known for his work in the computer domain.

In the lobby of the Harvard Computation Laboratory, prominently displayed, was a set of Babbage computer wheels in a small glass case, flanked by Aiken's copies of some of Babbage's books. This demonstration model, assembled by Charles Babbage's son Henry Prevost Babbage, was later mounted on a wooden base about 14 by 14 inches,

---

3. The Continental algorithm, devised by Leibniz and his followers, made use of the now-familiar $d$ (as in $ds/dt$ or $d^2x/dt^2$); the English algorithm, devised by Newton, made use of overdots (a single-dotted $x$ for $dx/dt$, a double-dotted $y$ for $d^2y/dt^2$, and so on).

with an engraved plate reading "Calculating Wheels, Designed By Charles Babbage, 1834." When Aiken first encountered these little wheels, they were not properly mounted. It was Aiken who had them placed on a handsome wooden base and who designed the metal plate identifying them. During the 1940s and the 1950s, Aiken used a slide showing these wheels in his public lectures.

The earliest expression I have found of Aiken's appreciation of Babbage's creative efforts occurs in the formal proposal submitted to IBM. In the introductory historical résumé, Aiken describes Babbage's ideas on machine calculation with clear admiration. In its final version (prepared by November 1937), this document begins with the statement of the "desire to economize time and mental effort in arithmetical computations; and to eliminate human liability to error," which is "probably as old as the science of arithmetic itself" and which "had led to the design of and construction of a variety of aids to calculation, beginning with groups of small objects, such as pebbles." Later, Aiken notes, "beads" were "mounted on wires fixed in a frame, as in the abacus." Then, after brief references to the devices of Napier, Pascal, Moreland, and Leibniz and a discussion of the slide rule, Aiken turns to Babbage and his two great machines.[4] This leads to mentions of desk calculators and the Hollerith machines. Aiken concludes the introductory section with the observation that many of Babbage's goals were realized in the daily operation of IBM machines.

Close examination of Aiken's proposal shows that at that time he did not have detailed and accurate knowledge of the purposes and principles of operation of Babbage's two proposed machines. He was aware that the later Babbage machine, the Analytical Engine, "pointed the way . . . to the punched-card-type of calculating machine" to the degree that "it was intended to use perforated cards for its control, similar to those used in the Jacquard loom." But Aiken was so far from comprehending Babbage's mature plans that all he said about the Analytical Engine was that it had been designed to be "of far higher powers" than the earlier Difference Engine and that it was "intended to evaluate any algebraic formulae by the method of differences." Nonetheless, his statements make clear that by 1937 he was familiar with some of the basic features of Babbage's Difference Engine and Analytical Engine.

4. He also mentions the difference engines built by Scheutz, Wiberg, and Grant.

The historical introduction in Aiken's proposal concludes with notes on the punched-card systems developed by Herman Hollerith and on IBM's recent commercial machines. In contrast to earlier devices, according to Aiken, Babbage's machines had "been designed strictly for application to scientific investigations."

The opening chapter of the Manual of Operation for the ASCC/Mark I (written by Grace Hopper under Aiken's direction and published in 1946) announces in several ways that Charles Babbage had a serious claim to priority in the realm of ideas about computing by machine. Facing page 1 was a plate showing Harvard's set of Babbage computing wheels. A quotation from Babbage's *Passages from the Life of a Philosopher* (1864) was used as an epigraph to chapter 1:

> If, unwarned by my example, any man shall undertake and shall succeed in really constructing an engine embodying in itself the whole of the executive department of mathematical analysis upon different principles or by simpler mechanical means, I have no fear of leaving my reputation in his charge, for he alone will be fully able to appreciate the nature of my efforts and the value of their results.

When Aiken "first came across these lines," according to Jeremy Bernstein, Aiken "felt that Babbage was addressing him personally from the past." "If Babbage had lived 75 years later," Aiken told Bernstein, "I would have been out of a job."

The "Historical Introduction" to the Manual of Operation—a much-expanded version of the introductory paragraphs of Aiken's 1937 proposal to IBM—not only deals with Babbage's Difference Engine and Analytical Engine but also traces the history of digital calculation from the invention of the abacus to IBM's punched-card machines. It concludes with a series of explanations for Babbage's "failure to complete either of his projects," which—according to Aiken—was not attributable to "a lack of understanding of the principles and purposes of the engines that he designed."

Shortly after the Manual of Operation was printed, Aiken and Hopper published a summary of its contents in the journal *Electrical Engineering*. Here too the precursorship of Babbage was stressed. Although none of Aiken's early writings make precise either what he learned from Babbage (e.g., ideas concerning control) or when he first encountered Babbage's work, Aiken's own publications give sanction to Brian Randell's statement that "he considered himself influenced by Babbage."

The intellectual bond between Babbage and Aiken was publicly emphasized by the presence of Richard H. Babbage at Harvard's 1947 Symposium on Large-Scale Digital Calculating Machinery. Babbage, a journalist from Canada, was a great-grandson of Charles Babbage. Attending the symposium on Aiken's express invitation, he contributed the opening technical paper, "The Work of Charles Babbage"; it consisted largely of selections from the published writings of his distinguished grandfather.

When Henry Tropp and I interviewed Aiken, one of my goals was to evaluate the extent of his indebtedness to Babbage. I asked him how and when he had come upon Babbage's ideas and how much they had influenced his thought, hoping to get answers to these questions: At what stage in the development of his own ideas had Aiken first heard of Babbage's "engines"? What were his sources of information, and how much did he actually learn about Babbage's work? Had he really been influenced in any major way by Babbage? Was it possible that he had merely adopted Babbage as a hero—a giant of the past who could give his work legitimacy?

Aiken was sitting on a couch in his living room. Above his head was a shelf containing bound copies of all the doctoral dissertations written under his direction at Harvard, a small silver model of Mark I, and some books. Among the latter I espied two copies of Babbage's autobiography, *Passages from the Life of a Philosopher*. Looking directly at the two books, I asked Aiken how he happened to have two copies. He took them down from the shelf and put them in my hands. I read aloud (into the tape recorder) the inscription written on the flyleaf of each. From the first copy I read

From one admirer of Babbage to another.
L. J. Comrie to Howard H. Aiken.
1946 March 8

Aiken's second copy of Babbage's book had once belonged to a Charles H. Pillemer and had been a gift "from W.M.C. February 1889." Underneath that inscription, however, was the name of a later owner: J. W. Bryce. Under Bryce's name was Aiken's. The provenance of these two books, with their inscriptions, informs us only that Aiken knew of Babbage's autobiography some time after he had worked out his plans for a calculating machine of his own. Clearly, these two books were gifts made long after Aiken had completed his proposal to IBM. Their presence on his shelf did not provide a key to the date of Aiken's first encounter with Babbage's ideas; they were only further testimony to Aiken's admiration for Babbage.

I then asked Aiken exactly how and at what point in his career he had encountered the work of his predecessors, and notably Babbage. The answer was immediate. While he was working on his thesis at Harvard, Aiken said, he needed to use cylindrical coordinates, and some problems arose in which there were differential equations in nonlinear terms. "Nonlinear differential equations are not the easiest to deal with on analog computers," Aiken explained, and for that reason "I began being interested in digital machinery." Eventually he turned up to two books in Harvard's Widener Library, both of which he read carefully. One was *Modern Instruments and Methods of Calculation: A Handbook of the Napier Tercentenary Exhibition*, edited by E. M. Horsburgh (1914). The other was *Calculating Machines and Instruments: Catalogue of the Collections in the Science Museum*, edited by D. Baxandall (1926). These "two books," he said, "provided me with an enormous amount of background information." Later he found Babbage's *Passages from the Life of a Philosopher.* With respect to these three books, Aiken remarked: "There's my education in computers, right there; this is the whole thing, everything I took out of a book." Had he been led to Babbage's autobiography by historical references in the other two books? "Not at all," Aiken replied. He hadn't encountered *Passages* until after his own ideas had been rather developed, he said, and his doing so was related to an "amusing incident."

When Aiken was still a graduate student (apparently in 1937), he "set forth" his "proposals" for a new kind of calculating machine. "The faculty [of the Harvard Physics Department] had a rather limited enthusiasm about what I wanted to do," he recalled, "if not a downright antagonism." After the "faculty had begun to make it rather clear" that they "had no interest" in the proposed machine, however, people began talking about it. One day the chairman of the Physics Department, Professor Frederick A. Saunders ("I guess it was"), mentioned Aiken's idea to Carmello Lanza, a technician in the physics laboratory. According to Aiken, Saunders reported to him the substance of Lanza's comments about the proposed calculating machine. In Aiken's words, Lanza just "couldn't see why in the world I wanted to do anything like this in the Physics Laboratory, because we already had such a machine and nobody used it." It is easy to imagine Aiken's astonishment on receiving this information. He told Tropp and me that he immediately sought out Lanza and "demanded to know where we had this machine." Lanza led him up into "the attic" of the old Research Laboratory of Physics (now the Lyman Laboratory). There, "sure enough" Aiken told us, were the wheels that Aiken later "put on display in the lobby of the Computer Laboratory." With them was a letter from Henry Prevost Babbage describing these wheels as parts of his father's proposed calculating engine. This was the first time Aiken "ever heard of Babbage," he said, and it was this experience that led him to look up Babbage in the library and to come across his autobiography. By that time, Aiken had already presented his first proposals for a calculator to the Harvard Physics Department.

In *Passages from the Life of a Philosopher,* chapter V ("Difference Engine No. 1"), chapter VI ("Statement relative to the Difference Engine, drawn up by the late Sir H. Nicholas from the Author's Papers"), chapter VII ("Difference Engine No. 2"), and chapter VIII ("Of the Analytical Engine") would have given Aiken a rather good idea of Babbage's machines.

It is one of the ironies of history that Aiken does not appear to have come across *Babbage's Calculating Engine,* edited by Babbage's son Henry and published in 1889. Accurately described by its subtitle, *Collection of Papers Relating to Them, Their History, and Construction,* that volume contained L. F. Menabrea's celebrated "Sketch of the Analytical Engine" (1843), translated and annotated by Byron's daughter, Ada, Countess of Lovelace. As Brian Randell comments, if Aiken had read this volume he would not have been ignorant of Ada's "clear description of the programming and the theoretical possibilities of the machine, and of the importance of conditional branching."[5] Randell concludes that if Aiken "had known more about the Analytical Engine, he would surely have included conditional branch facilities in the original design of the Mark I." Most of the retrospective discussions in print concerning Mark I generally stress this limitation, even pointing out that only later was Mark I provided with a supplementary unit to provide "subsidiary sequence controls." There is, however, some confusion with respect to this point. It is true that a permanent unit for branching was added later. However, Robert Campbell recalls, "We started programming branch operations in only a few weeks" after the

---

5. Grace Hopper has recalled that "almost the first day [she] met a computer [in 1944]," she "met Babbage." She relates that "Commander Aiken had a copy of Babbage's book, and at intervals advised us to read sections of it." She did not, however, "meet Lovelace's work until ten or fifteen years later."

machine was put into service at Harvard, using only "a minor change in wiring."

I had one last question about Babbage to put to Aiken: Had he ever actually gone to the Science Museum at South Kensington to see any of the Babbage machines? "I went to the South Kensington Museum after the war," he replied. "I stayed with the Comries, and Comrie made arrangements for the museum to be opened up so that I could get to see the Analytical Engine." The museum was "still in a shambles," but Aiken did get to see the "mill" of the Analytical Engine (corresponding in its functional role to the CPU of a modern computer)—"and it was," he said, "the greatest disappointment of my life."

Aiken told Tropp and me that he first met L. J. Comrie in the early days of operation of Mark I, shortly after the publication of the first volume of tables. "Comrie," Aiken related, "got one of the first copies and he differenced those tables to see if he could find some errors." When I suggested that Comrie may have been looking at third differences, Aiken said "Oh, no! Seventh or eighth differences, because that table was very coarse. . . . If it had a fine interval, it would have been a whole library. So he had to go to very high-order differences, which he did, looking for errors in that book." Almost grudgingly, Comrie admitted "that he had found no errors." Aiken continued: "Well, that was the beginning of my association with [Comrie]." Aiken recalled their "very pleasant association" over many years—an association that, though "very constructive," always had, on Comrie's part, "this theme of envy that what we were doing hadn't been done in England." Comrie, according to Aiken, could not forgive his shortsighted countrymen for "never [having] fostered any development of this kind." Aiken recalled Comrie's published judgment that "this dark age in computing machinery, that lasted 100 years, was due to the colossal failure of Charles Babbage," who "spent all that money and nothing came of it." As a result, "academic people simply threw up their hands. . . . This was the thesis of Comrie." And indeed, Comrie's review of the Manual of Operation in *Nature* began: "The black mark earned by the government of the day more than a hundred years ago for its failure to see Charles Babbage's difference engine brought to a successful conclusion has still to be wiped out. It is not too much to say that it cost Britain the leading place in the art of mechanical computing." Linking the ASCC/Mark I with Babbage's invention, Comrie concluded that, although the American machine in "its physical form has the benefit of twentieth-century engineering and mass-production methods," it was "in principle" basically "a realization of Babbage's project."

The last big question here is whether Babbage's machines had any influence on the design of the ASCC/Mark I.

Although there seems to be no way to learn the actual date on which Aiken approached the Physics Department and encountered the Babbage wheels, there are some convincing reasons why this event would have occurred in 1937, at about the time when Aiken had written up his formal proposal for IBM and would have had real hope that his machine might be built. The chief reason for this date is that Aiken would surely not have approached the Physics Department with a request for space if he didn't have a real prospect of his dream machine's becoming a reality.

When he was writing his proposal, Aiken's knowledge of Babbage's machines, particularly the Analytical Engine, was severely limited; it was the kind of information gleaned from meager secondary sources rather than from a reading of Babbage's own writings about his machines, as in his autobiography. As has just been noted, Aiken mentioned the Horsburgh and Baxandall volumes during the 1973 interview. That those two books, and not Babbage's autobiography, provided Aiken with his knowledge of Babbage's machines in 1937 is easily documented by a comparison of what they say about Babbage and his machines and what Aiken wrote in his proposal to IBM.

A close reading leaves no doubt that many of Aiken's statements about Babbage's two machines are taken directly from Baxandall's *Calculating Machines and Instruments* (1926), which is basically an annotated and illustrated catalogue of the collection in the Science Museum at South Kensington. *Calculating Machines and Instruments* gives basic information (maker, date of manufacture, donor, date of acquisition) concerning 239 numbered machines or devices, classified under such general rubrics as "calculating machines," "difference and analytical engines," "slide rules," "instruments for solving equations," "linear integrators, planimeters, integrometers," "integraphs," and "harmonic analysers and integraphs." Each section contains a brief introduction, a page or two in length. There are thirteen plates, one showing the Babbage Difference Engine and another the Scheutz Difference Engine; both contain clear reproductions of wheelwork, with numbers visible.

Baxandall's introduction to the section on difference and analytical engines is primarily devoted to an explanation of the method of differences and the chief operations of the Difference Engine in producing tables. The Analytical Engine is treated much more summarily. Baxandall mentions, however, that the machine's "various operations" would be "controlled automatically on the Jacquard principle."

An example of Aiken's use of Baxandall's text occurs in Aiken's observation (in his proposal) that, after Babbage gave up his work on the difference engine,

. . . other difference engines [were] constructed and designed by Martin Wiberg in Sweden, G. B. Grant in the United States, Léon Bollée in France, and Perry Ludgate in Ireland, which, however, were never constructed.

Compare this with Baxandall's statement:

Other difference engines were designed and made by Martin Wiberg (1863) in Sweden, G. B. Grant in the United States; others were designed by Léon Bollée in France, and Percy Ludgate in Ireland, which, however, were never constructed.

Elsewhere, Aiken even followed Baxandall in Anglicizing Georg Scheutz's first name to "George."

Aiken said very little in his proposal about the actual architecture or mode of operation of the Analytical Engine, only that it was "intended to evaluate any algebraic formulae by the method of differences." The second part of this statement seems attributable to a hasty reading of Baxandall, who wrote that by using "this engine, which would embody features of the difference engine, any mathematical formula was to be evaluated." In Horsburgh's book this same set of words is used in relation to an "algebraic formula" rather than the more general "mathematical formula." Another coincidence of phrases may be seen again in Aiken's comment that, after "abandoning the difference engine," Babbage "devoted his energy to the design and construction of an analytical engine of far higher powers than the difference engine." This too is taken almost word for word from Baxandall:

Babbage devoted his whole energies and resources to the design and construction of an "analytical engine" of far higher powers than the difference engine.

Baxandall's statement that Babbage's Analytical Engine "proved to be too ambitious" found an echo in Aiken's conclusion that the same machine was "too ambitious" for its time.

Additionally, Baxandall did not stress the fact that Babbage intended to have his machine print out the results directly in molds from which stereotype blocks could be made for printing (thus eliminating errors due to transcription, typesetting, proofreading, and printing). Aiken did not mention this feature of the Babbage engines; rather, he presented it as his own innovation.

Horsburgh's *Modern Instruments and Methods of Calculation* was primarily a collection of short descriptive articles on mathematical tables, various types of calculating machines in production and use before World War I, the abacus, slide rules, nomograms, and various "other mathematical laboratory instruments" (including integraphs, integrometers, planimeters, and equation solvers). Its basis was an exhibition celebrating the tercentenary of the publication of Napier's classic work on logarithms. The opening sections of the Horsburgh volume deal with Napier's invention and the "antiquarian" part of the exhibition. From this volume Aiken would have learned of many early machines, especially those of the late nineteenth century and the early twentieth.

In one article—"Automatic Calculating Machines," by Percy E. Ludgate—Aiken would have encountered the challenging assertion that "the true automatic calculating machine belongs to a possible rather than an actual class." Most of Ludgate's article is devoted to the Analytical Engine. "This engine," writes Ludgate, "was to be capable of evaluating algebraic formulas, of which a numerical solution is possible, for any given value of the variables." This sentence was modified by Baxandall so as to refer to a "design . . . for calculating and printing the numerical values of any algebraic formula, where a numerical solution is possible, from given values of the variables." As has been noted above, this appears in Aiken's proposal as the intention "to evaluate any algebraic formulae."

The conclusion to which we are led is that Aiken initially learned about the work of Babbage primarily through the writings of Baxandall, in part supplemented by Ludgate's brief article. It was from these sources that he derived the summary (and somewhat erroneous) statements about Babbage's machines found in his 1937 proposal to IBM. That Aiken was generally ignorant of Babbage's machines explains the otherwise puzzling fact that, though Aiken continually expressed admiration for his illustrious predecessor, the architecture of the ASCC/Mark I was little influenced by Babbage's architecture.

Reviewing all the evidence, and taking Aiken's oral testimony at its face value, one would have to conclude that Babbage did not play a major role in the development of Aiken's ideas about machine architecture. By his own account, Aiken came across detailed information on Babbage's Difference Engine and Analytical Engine only after he had completed a general plan for a digital super-calculator of his own and proposed it to Harvard's Physics Department.

There is no evidence that Aiken declared indebtedness to Babbage for any particular features of his own inventions. Rather, it seems, Aiken praised Babbage in order to enhance his own stature. By lauding Babbage's intellectual prowess and conceptual achievements, Aiken seems to have been saying in effect that in his bold pioneering effort he, Howard Aiken, was not alone. He was suggesting that he was in some sense like Babbage: a radical inventor whose stature was not fully appreciated by his contemporaries, a member of a small group of great men (including Pascal and Leibniz) who had pioneered the art of mechanical computation. By declaring his personal admiration for Babbage, and paying tribute to other illustrious predecessors, Aiken was not immodestly staking his own claim to a place in history.

It also seems that, as he got deeper and deeper into computing, Aiken—fatherless since his early teens—adopted Babbage as a kind of father figure, somewhat as, in earlier days, he had done with the high school teacher who assured him of an education, with Edward Bennett at the University of Wisconsin, and with E. L. Chaffee at Harvard.

Aiken's sense of his own importance was manifested in a number of ways. For example, he documented every stage in the development of Mark II, Mark III, and Mark IV, and all the events at Harvard connected with Mark I. There was always a photographer on hand to record every stage of the planning and construction of these machines and even of the buildings in which they were being designed, assembled, or put into operation. The contribution of every member of the staff—including secretaries and maintenance and support staff—was recorded and listed at the front of each volume of the *Annals of the Computation Laboratory*.[6]

Not surprisingly, Aiken differed from many of his contemporary pure and applied scientists in highly valuing the history of science as a serious academic discipline. Perhaps he thought that his own contributions might someday be subjects of historical investigation.

6. Though Aiken is always listed as "Chief" or as "Officer in Charge," he never declared himself to be the sole author or even a primary author.

# 9
## Planning and Beginning the Construction of the Machine

Because almost all of Aiken's early papers were lost when he was called to active duty in the Navy, it is difficult to document each step in the planning and actual construction of the ASCC/Mark I. However, the chronology of the early events can be established from correspondence, transcripts of interviews, and other documents in the IBM archives,[1] supplemented by the materials in the Harvard University Archives.

In the late 1930s, when Aiken made contact with IBM, James Bryce, Clair Lake, Francis Hamilton, and Benjamin Durfee were old-timers in the company, their association going back to the days of the Computing-Tabulating-Recording Company. By 1924, when Thomas Watson renamed CTR the International Business Machines Corporation, punched-card apparatus for commercial purposes was becoming the company's main product line. Bryce, as has been mentioned, joined CTR as a draftsman and engineer in 1900 at the age of 20. Lake joined CTR in 1915. IBM's historians note that Lake's "credentials were based on performance, not on education." After completing the eighth grade, he had gone to a manual training school. In 1915 he was "an inventor with experience in automotive design." His first assignment at CTR was to design a printer to be used in conjunction with tabulators. Among his important contributions was a method of simplifying "the adding mechanisms of the tabulator, a step that eliminated many relays and thereby gave him freedom to bring the plugboard from the back of the machine to a more accessible location at the front."

1. The IBM archives contain transcripts of interviews with Hamilton and with Durfee, who were in charge of the day-to-day work of design and construction of the new machine. These interviews were conducted by Larry Saphire as part of an IBM history program. One important source of information is a detailed 15-page historical review written by Hamilton.

At the time of Aiken's first contact with IBM, Frank Hamilton, who had joined the staff of CTR in 1923 as a draftsman, was Lake's assistant. Hamilton was a greatly talented but modest man. In the IBM interview conducted by Larry Saphire, Hamilton admitted that he "didn't know anything about computers at all" and that "he didn't know what those problems were" that the ASCC was designed to solve. Again and again in the course of the interview, he made it plain that he didn't know any mathematics other than the simplest algebra. "I had no mathematical education at all," he said, and "neither did Lake and neither did Durfee." After working on the ASCC/Mark I, Hamilton became the chief engineer in charge of construction of IBM's Selective Sequence Electronic Calculator (SSEC), a machine intended by Watson to be better than Aiken's. Hamilton—later an important figure in IBM's Magnetic Drum Calculator project—eventually became manager of IBM's Endicott laboratory.

Ben Durfee joined CTR in 1917 and was sent to "training school" in 1918. His early reputation was "for diligence in servicing tabulators" ("checking machine adjustments, oiling and cleaning, and replacing worn parts"). Durfee spent the period 1924–1935 in Europe working on problems that arose in connection with the use of IBM machines. Durfee did most of the day-to-day work of assembling the ASCC/Mark I and wiring the circuits. Like Hamilton, he was a skilled mechanic and a clever inventor; however, like Lake and Hamilton, he lacked an advanced education and a knowledge of higher mathematics. In the course of one IBM interview, Durfee said he "had never been too good at mathematics."

On 3 November 1937, Aiken wrote to Bryce asking for an appointment, remarking that "Professor Brown of the Graduate School of Business Administration" had "discussed with you my interest in automatic calculating machinery for use in computing physical problems." Bryce replied the next day, suggesting two possible dates; Aiken made his choice and responded on 6 November. On 10 November, Aiken met with Bryce, who then became interested in the possibility of constructing a new type of machine to Aiken's specifications. Bryce, according to Hamilton, had "talked with Professor Brown at Harvard concerning the proposed development of the Calculating Machine." Acting on Brown's recommendation and his own sound judgment, Bryce decided to encourage Aiken.

Bryce was aware, however, that Aiken knew very little about the actual functioning of IBM's machines or about the IBM machines being used by Wallace Eckert and his associates at the computation center established by IBM at Columbia University. Accordingly, Bryce arranged for Aiken to visit Eckert and to see the Columbia installation. Apparently he wanted Aiken to find out whether the Columbia facility would be adequate for dealing with the kind of problems Aiken had in mind. When Aiken reported that his needs far exceeded the potentialities of the Columbia machines and required an entirely new type of machine, Bryce decided that Aiken had better learn something about IBM's products, many of whose components would become elements of Aiken's machine. Bryce apparently got in touch with Theodore Brown to make arrangements.

And so once again Ted Brown was the intermediary between Aiken and IBM. Brown put Aiken in contact with H. E. Pym, manager of IBM's Boston office. The course of Aiken's learning about the "actual working" of IBM machines can be traced through the letters exchanged by Pym and Bryce during December 1937 and January 1938—carbon copies of which were always sent to W. Wallace McDowell at IBM's Endicott facility. McDowell, a graduate of MIT, was one of the first men with a college degree in engineering to be hired by IBM. Upon joining up in 1930 he was assigned to attend what was know as the "sales school," but he was later transferred to Endicott.

Aiken began his IBM training by attending classes at the Customers' Supervisors School in Boston. After that, he went out on jobs with IBM servicemen. Late in January 1938, Pym asked Bryce by teletype: "Is it okay to explain factorial multiplier for proposed dividing machine to Aiken of Harvard University?" Bryce replied: "It is all right to explain to Aiken the factorial multiplier. You can tell Aiken the dividing machine is under way but do not explain any details to him." This was Aiken's first encounter with concerns about patents, with problems of secrecy and disclosure, and with the safeguarding of proprietary interests in the area of machines.

In February, according to a report by McDowell to Bryce, Aiken spent four days at Endicott "working with Mr. Hamilton and Mr. Durfee." Their discussions centered on "the solving of algebraic equations." A "rough estimate" of "the cost of our standard equipment to do the job" was to be prepared. At this time, all concerned with the project assumed that IBM would supply expertise and parts, but that the wiring and the construction of the machine would be done at Harvard. McDowell's report stated specifically his understanding that "all wiring and building the special devices would be carried out at

Harvard" but that IBM "would supply our standard units such as counters and relays." McDowell expressed some concern about drawing up an agreement with Harvard. He wanted to be certain about protecting the "rights which we would have in these new circuits or new methods which might be developed" and "the sole use of these things from a commercial standpoint." He then mentioned that Aiken had suggested "a means for division which may be patentable, and be of value to us." (This appears to refer to the possible use of the Newton-Raphson method of approximation.) He stressed that someone from IBM would have to go to Cambridge "when they start on the wiring diagrams," which "will probably be a fairly long job."

According to Hamilton's written report, the four days Aiken spent in Endicott were 31 January and 1–3 February 1938. Aiken "briefly outlined the purpose of his visit, which was to discuss the possibilities of designing a machine of a very complex nature for solving equations pertaining to the ionosphere in relation to work on television, and also concerning equations pertaining to astronomy." Clearly, Aiken tried to show the purpose of his proposed machine by linking it to television and to other practical applications within the range of understanding of IBM engineers not trained in mathematics and science at the college level.

According to Hamilton, "Mr. Lake turned Mr. Aiken over to Mr. Hamilton for further detailed discussion concerning the machine." The "entire time that he was at the Laboratory during this visit was devoted to obtaining a more thorough idea of the requirements of the machine." As this entry makes clear, Lake had the overall responsibility for the project, while Hamilton was the engineer primarily responsible for the actual design and construction of the machine. Durfee was charged with making the circuits and assembling the parts.

The importance given by Bryce to Aiken's proposal can be gauged by the fact that when Aiken first visited IBM's laboratory in Endicott, New York, in February 1938, his official host was Wallace McDowell, who—as has been mentioned—held the post of assistant to the vice-president in charge of engineering; he was the first manager of the Endicott laboratory to be given that official title. Bryce had arranged for McDowell to introduce Aiken to Clair Lake, one of IBM's senior inventors or engineers who—in that capacity—had a staff consisting of an assistant engineer plus a group of designers, model makers, and draftsmen. As one of IBM's senior engineers, he was entitled to occupy one of the four large corner-offices in the laboratory. His high position can also be seen in the fact that he reported directly to the president of IBM, T. J. Watson. He was responsible for many important innovations that strongly influenced IBM technology, including the 80-column punched card with its rectangular-hole format that was a staple for most IBM machines after 1928. Other notable improvements for which he was known include new types of rotary counters, circuit breakers, and plug-in relays—all of which "profoundly influenced the design of IBM products."

During Aiken's first meetings with the IBM engineers, only the four arithmetic functions were discussed: adding, subtracting, multiplying. and dividing. The "more complicated functions, such as logarithms, sines, tangents, and interpolation" were, according to Hamilton, "to be discussed later." Aiken expressed his "opinion that the control of the machine should be obtained from the use of a tape." Aiken did not "disclose" any "definite ideas as to what the tape should be like either for size or for coding of the holes," nor did he "disclose any idea as to the means to be used for operating this tape."

Hamilton stressed that Aiken knew what he wanted the machine to be able to achieve but did not have "definite ideas" about its functioning. Since Hamilton wrote his report after the machine's dedication (it is dated 21 August 1944, three days after the ceremonies), he was at pains to assert Aiken's innocence of technical details and the importance of the IBM engineers in devising functional systems to make the machine work.

One important topic discussed at the early meetings was the method to be used for storing constants. Hamilton first suggested "the use of the IT cross-connecting program panel," but this was abandoned because "to change the setup from one problem to another" would require too great an expenditure of time. Hamilton then proposed "the use of rotary switches," which would be efficient "from an operational standpoint" and which would have the advantages of being "more readily changed" and "more easily read." Hamilton noted that this "idea was maintained throughout all the planning of the machine" and that the machine "is now equipped with these switches."

Hamilton also recorded the discussions with Aiken concerning the methods for division. At the time, IBM machines did not have an adequate capacity for division. Aiken suggested a method of division by computing reciprocals of the divisor. This is apparently the process,

78   Chapter 9

based on the Newton-Raphson method, mentioned earlier in Bryce's letter. Aiken's proposal was rejected, however, because Bryce had invented a different mode of division that used fewer machine cycles than Aiken's suggested method. It was Bryce's system of division, then being developed at IBM, that H. E. Pym had been uncertain about allowing Aiken to learn of. Hamilton described Bryce's method succinctly as "based on all nine multiples of the divisor being stored in the machine and chosen dependent on the relation of their value and the value of the dividend." The "chosen value would then be subtracted from the dividend." Hamilton observed that this "method of dividing is the one that is employed on the machine at the present time."

Hamilton also recorded an ulterior motive: here was "an excellent opportunity to incorporate" the Bryce divider into a working machine in order to "prove [that is, test] the value of the increased speed obtained by its use." The giant new calculator would provide a testing ground for new devices being developed at IBM. (The Bryce divider was later removed from the machine by Aiken, who devised better methods of dividing.)

Shortly after Aiken returned to Cambridge from his four-day visit to Endicott, Harry Mimno wrote a report to Harvard's President Conant on the state of affairs with respect to the Aiken project. Clearly, Aiken could not enter into serious negotiations with IBM or any other outside source without the approval of the Harvard administration.

The first section of Mimno's report (dated 7 February 1938) is a reminder to Conant of the contents of a previous report, written in the late spring of 1937, informing him about the steps Aiken was taking to turn his dream of a super-calculator into reality.

The second section brings Conant up to date on Aiken's activities. Mimno reminds Conant that "Mr. Howard Aiken, of the Cruft staff, had developed plans for an automatic computing machine designed for the solution of fundamental scientific problems." It has become evident that this is apt to "develop into a practical interdepartmental project." The "success of the venture" has been "made probable" by "the fact that the proposed machine employs a considerable number of standard business machines as component parts." (The IBM engineers thought the new machine would be a computing "plant" made up of several machines or components, linked together.) In February 1938, IBM had not firmly committed itself as the sole sponsor of the enterprise. However, it had been "unofficially suggested by responsible members of the International Business Machines Company[2] that their organization may offer to supply all of the standard component parts without charge." Mimno reckons that this will come to "at least $50,000 (company cost)." He reminds Conant that Harvard's "own staff would assemble the machine." Harvard will "have to construct the special central control mechanism which is to govern the operation of the numerous standard parts." Mimno hopes that "a considerable part of this control switchgear" might be obtained "in a semi-finished state by soliciting contributions of control devices used by the Western Union Company in conventional Teletype installations or by the use of automatic telephone parts." Harvard will have to provide "only a few thousand dollars" for "incidental expenses." The "cash contribution from University funds would be relatively small in comparison with the magnitude of the project." In the event, Harvard did not have to apply to Western Union or to the telephone company, since IBM shouldered the full financial burden of parts and construction.

The third section of Mimno's report to Conant is perhaps the most interesting. Here Mimno sets forth some important potential uses of the new calculator in furthering the research of Harvard faculty members. After all, the primary justification for Harvard's support of Aiken's project is that the new machine will be of real service to science and engineering. Evidently Mimno had spoken with various members of the Harvard faculty whose research involves extensive calculation. Some potential applications of the new super-calculator, he notes, include astronomical problems assembled by Theodore Sterne and Harlow Shapley, "important applications in chemistry" proposed by Albert Sprague Coolidge (then performing extensive calculations arising from the quantum theory of hydrogen), and "problems in pure mathematics . . . mentioned by Dean [George] Birkhoff." Professor Theodore H. Brown of the Graduate School of Business Administration foresees "probable applications in statistical researches on business problems." Professors E. C. Kemble, Wendell H. Furry, and J. H. Van Vleck "are interested in difficult numerical problems in Atomic Physics." Dean Westergaard and an associate "have specific problems in

2. In referring to the "International Business Machines Company" rather than using the correct name, International Business Machines Corporation, Mimno made the same error that Aiken had made in his original proposal. In those days, quite obviously, members of the Harvard faculty had not had much experience in dealing with industry and were not aware of the importance of getting such names right.

80  *Chapter 9*

Applied Mechanics." The "opportunities," Mimno concludes, "are practically limitless." Mimno stresses to President Conant that "all . . . the scientific problems are of major importance" and that many of them would require "an absolutely prohibitive amount of human labor if attacked by conventional methods."

The penultimate section of Mimno's report concerns patents and profits. In consideration of such a "major contribution to scientific progress by the International Business Machines Company," the Harvard staff "would be entirely willing to allow the company to patent any incidental improvements suggested by this novel application of their devices."

Mimno concludes by emphasizing a point of view, held generally at IBM, that a senior official at IBM soon expressed explicitly: "It is altogether unlikely that the machine as a whole will ever have commercial value." Being "designed exclusively for scientific applications," it is "much too expensive for general distribution." Mimno does, however, express his belief that "members of the manufacturer's own research staff will undoubtedly be stimulated by close contact with persons who are attacking refreshingly new problems in the design of computing apparatus." Furthermore, "company engineers have already indicated that they would thoroughly enjoy an opportunity to collaborate on such an experiment." "Naturally," Mimno concludes, "the company will also derive added prestige through the use of its apparatus in a project which is likely to arouse world-wide interest when successfully completed."

Hamilton's historical memorandum reveals in detail how the working design of the new machine was developed. For example, it was the "decision was made by Mr. Hamilton and Mr. Durfee that all adding counters, switches, card feeds, etc. be connected to a single buss of 24 columns for purposes of transferring values"—a decision, Hamilton proudly observes, that greatly reduced "the amount of connections and wiring in the machine."[3]

McDowell proposed "that photocells be employed as a means of sensing the code punching in the control tape." Aiken wanted "the use of teletype" considered. When Aiken designed and built Mark II, he actually did use teletype apparatus.

---

3. The choice of 23 digits for calculations and one place to indicate whether the number is positive or negative is discussed in chapter 18 below. As was noted in chapter 7, Aiken's original proposal contained a short statement concerning the reason for the use of 23 digits.

It was decided by Lake and Hamilton that IBM card stock would be used and that the perforations would be sensed by "contact fingers" as in standard IBM machines. Here again Hamilton stresses that Aiken's technical suggestions were less practical than the proposals of the IBM engineers. Hamilton mentions the agreement that the final place (the 24th) would be reserved for indicating algebraic sign (a 9 for plus and a 0 for minus) and that subtraction would be performed by complement addition.

The location of the decimal point was discussed, and "Mr. Aiken established the principle that all values throughout the solution of a problem should have the decimal point in the same position." Means "for attaining this principle were incorporated into the machine circuit design by Mr. Hamilton and Mr. Durfee." Here again Hamilton's effort to make a clear distinction between Aiken's suggestions and those of the IBM engineers is evident.

Aiken would set forth a "principle"; then Hamilton and Durfee would devise a combination of machine components that would operate to achieve the desired end. This division of labor could not help but generate disagreement. Aiken believed that he had initiated the train of ideas leading to a part of the machine and, indeed, to the conception of the machine as a whole. He thus conceived himself to have been "the inventor" or at least the primary inventor. From the IBM viewpoint, however, most (if not all or almost all) of Aiken's "practical" suggestions had been rejected. Aiken had, indeed, told the IBM engineers of his needs, but the IBM engineers had then devised operational units to achieve Aiken's rather abstract goals. They "invented" ways of doing what Aiken required. From their point of view, they had actually invented and constructed a machine to perform the operations Aiken had specified. Without their experience and inventive skill, Aiken's ideas would never have been translated into a working device.

As has been mentioned, a short time before Mimno wrote his report to President Conant, in February 1938, Wallace McDowell expressed to James Bryce his concern about proceeding without some formal agreement between Harvard and IBM. Bryce's reply was to write to Conant in order to arrange a meeting with him to determine the "financial procedure." In his letter, dated 4 February and thus written just after Aiken's four-day visit to Endicott, Bryce explains why he believes that Aiken's method for performing division, though "ingenious," is not the best system for a machine. Evidently, Ted Brown had assured Bryce that Harvard intended to "turn over to us any inventions that may grow out of the work they do with us in connection with the machine." At this stage, the cost of machine components to be furnished by IBM was still estimated at be about $15,000—paltry in comparison with the eventual cost of about $200,000.

In the event, in March, John George Phillips of IBM went to Harvard at "the request of F. W. Nichol." Nichol was Thomas Watson's longtime confidential secretary and personal assistant[4]; Phillips was Watson's "private financial secretary, keeping track of most of the business affairs of his entire family."[5] Phillips describes this visit in a memorandum dated 15 March 1938. He met a number of members of the Harvard faculty at a dinner meeting in Cambridge. Present were Harald Westergaard (dean of Harvard's Graduate School of Engineering), Theodore Sterne (a mathematical astronomer, representing the Harvard Observatory in place of Harlow Shapley, who could not attend), E. V. Huntington (a mathematician, a specialist in statistics), George Birkhoff (Dean of the Faculty and one of the world's leading mathematicians), P. W. Bridgman (a physicist famed for his work in high-pressure phenomena), E. C. Kemble (one of America's foremost theoretical physicists, a great teacher in such realms as mathematical physics and quantum mechanics), Howard Aiken, and Ted Brown (representing the Graduate School of Business Administration). Phillips found "a great deal of their technical conversation" to be "over [his] head"; however, "they reduced enough of it to the language of the layman to make it a very interesting talk."

At the dinner, according to Phillips, Aiken expressed belief in the feasibility of developing a prototype of the proposed calculating machine from standard parts of IBM accounting machines (with a "certain amount of additional work and wiring"). In this way, it could be shown how calculations in astronomy and physics could actually be performed. Such calculations then required "weeks, months or even years."

Phillips said that Aiken's machine did not have any "commercial application." This statement is a plain expression that at this time IBM

---

4. Nichol had been Watson's personal secretary at National Cash Register. In 1935 he became general manager of IBM. His involvement in the calculator project is an indication that the proposed machine had been brought to Watson's attention and that Watson had approved continued exploration of the possibility of IBM's building the machine.

5. Eventually, Phillips became vice-president and chairman of the board of directors of IBM.

("a commercial company") did not see the new machine as the beginning of a new product line. IBM's support for this project was based on Watson's eagerness to forge ties with the scientific and educational communities.[6]

But even though there was no hope of financial gain, IBM could reasonably expect certain real benefits from the joint enterprise with Harvard. This expectation accounts for a significant part of the later unpleasantness in relations between Aiken and Watson and between Harvard and IBM. In any case, it should be stressed that IBM did not support the Aiken-Harvard project because of any hope that the resulting machine would inaugurate a new product line or would have any business or commercial application.[7]

At this time, there seems to have been no further doubt that IBM would underwrite the whole construction of the machine. Bryce and other at IBM still believed, however, that the "wiring and assembly" would be done in the Harvard machine shops, while IBM would furnish standard parts and "cooperate in the cost of panels and other materials." The cost estimate had now risen from $15,000 to between $50,000 and $75,000, half of this sum being for parts.

On 22 March 1938, Bryce wrote to Lake to find out whether it was not time for IBM to "provide for an appropriation to start work on the Calculating Machine." Evidently, Lake and Hamilton had concluded that the Harvard staff did not have sufficient skill to construct the proposed machine out of IBM parts. On 24 March, Bryce informed McDowell that he was "very skeptical as to whether or not the men at Harvard would be able, or have enough knowledge, to mount such things as counters so that we could be sure that they would operate properly." Aiken evidently agreed, and the responsibility for all aspects of design and construction was assigned to IBM's engineers at Endicott.

Throughout March and April of 1938, Aiken furnished lists of constants needed for the machine. On 16 April, Bryce sent Hamilton a statement of the arrangement of various panels for the machine—an

6. He had already established the computing center at Columbia University.
7. The hoped-for benefits to IBM from an association with Harvard should be restated here because of the importance of this issue. They included the possibility of using the eventual Harvard machine as a testing device for new IBM technology and, perhaps most important, the use of the Harvard contact to evaluate the possible uses of IBM machines in science and engineering. And surely IBM would benefit from the ideas of Harvard's faculty and advanced students.

arrangement, he said, "that both Mr. Aiken and I feel . . . would be practical and satisfactory." Now that the specifications of the layout had been determined, Bryce wrote, he hoped that Hamilton "at last" could "get started." (In his historical memorandum, Hamilton referred to this letter from Bryce as having been written on 16 May, described the arrangement of the panels as "tentative," and added that the number of counter positions required was only "approximate.") On 19 May, Hamilton received a "work order" making funds available for a temporary layout which would show how the several units of the calculator were to be arranged. For some reason, work on the actual layout of the panel did not begin until 17 August. This first stage of work was completed on 12 September 1938, at which time the total cost of the machine had risen to an estimated $100,000, the sum on which Watson's final approval would be based. Not until February of 1939, however, did Watson actually give that approval. In May, an initial appropriation of $15,000 was made so that construction could begin in earnest. At that time it was officially confirmed that Clair Lake was to be the engineer in charge, with Hamilton and Durfee responsible for day-to-day design and construction. On 22 May, Aiken met with Lake, Hamilton, and Durfee in Bryce's office. At long last the project was really underway.

Aiken now began to make regular visits to Endicott, discussing such questions of a means of keeping "individual synchronous motors" running together "in time."

In the meantime, on 18 January 1939, James Bryce had sent to Harvard a memorandum stating that IBM and Harvard agreed to cooperate on producing "an automatically operated assembly of calculating machines." This curious wording is an indication that what was to be built was so complex that it could not be classified simply as a single IBM machine. (At the time, "an IBM machine" meant a typewriter, a collator, a sorter, an adding machine, an accounting machine, or an electromechanical business machine.)

The next three paragraphs of the agreement mention the types of problems the new calculator would be able to solve. (These paragraphs were taken, almost word for word, from the proposal Aiken had submitted in the autumn of 1937.) It was then specified that construction be based on "standard accounting machine parts so far as possible." Then the agreement discussed the proposed machine as "a series of computing machines, flexibly coupled together in a large number of combinations." It was "proposed to erect the various parts" of this complex of machines in the "building where they will be used."

The agreement specified that almost all the actual design and construction work was to be done by IBM, but there was still some hope that some of the "special mechanisms for . . . control of the machine" could be produced "at Harvard from designs made by IBM." The cost of the standard parts was now estimated at approximately $15,000, the total cost (including planning) at $75,000—$100,000. The time allotted for construction was 2 years.

The final paragraph is especially interesting in the light of later developments. It was noted that the "machine would be used by scientists" and that the results would "be published widely." It was expected that "naturally there will flow from this a considerable amount of publicity for IBM, as well as Harvard University." This publicity would follow from "the practice of scientific people to give credit for their sources of information or facilities supplied." This concluding statement emphasizes the IBM's recognition that, although the new machine might have some patentable aspects, its construction was not a commercial enterprise but a contribution to science. IBM's primary reward for a considerable expenditure of money, time, and parts was to be favorable publicity in the scientific community.

By 31 March 1939, the final agreement had been drawn up and signed. IBM agreed in §1 "to construct for Harvard an automatic computing plant comprising machines for automatically carrying out a series of mathematical computations adaptable for the solution of problems in scientific fields." (Note that now the complex machine is being called "an automatic computing plant" that comprises "machines.") Harvard agreed in §2 to furnish the structural foundation "without charge" and in §3 to appoint "certain members of the faculty or staff or student body" to cooperate with IBM's "engineering and research divisions" in the design and the testing. It was agreed in §4 that all Harvard personnel assigned to the project would sign a standard "non-disclosure" agreement to protect IBM's proprietary technical and inventive rights,[8] and in §5 that IBM would receive no compensation and would make no charges to Harvard. The finished "plant" was to become "the property of Harvard."

A letter from Dean Westergaard confirmed the "understanding that the computing plant will be for the use of Harvard in scientific fields and that no commercial use will be made of it by Harvard." On 10

---

8. This was to be the same agreement IBM used for all its employees, consultants, and associates.

May 1939, about a year and a half after Aiken's first approach to IBM, James Wares Bryce wrote Aiken that all the papers had been signed and that he was now "engaged in getting an appropriation put through." As soon as this was done, Bryce would "issue the shop orders" and "begin the actual work of designing and constructing the calculating machine."

# 10
## How to Perform Multiplication and Division by Machine

The design of the ASCC/Mark I embodied many of the numerical practices used in IBM's tabulators and accounting machines of the 1930s and the early 1940s. The business machines of the 1930s were based on digitally incrementing accumulators, which were essentially adders. Since the machines could run only in the forward direction, subtraction was performed by complement addition, just as in Pascal's calculators of the early seventeenth century. Either a means of converting numbers into their complements had to be incorporated in the machine or tables of complements had to be provided. Negative numbers (as in the case of business losses) would appear as complements, and those would have to be converted.

The earliest electromechanical business machines performed multiplication in the same way as simple desk calculators: by "over-and-under addition," which was essentially a way to reproduce in mechanical operations the individual processes used in long multiplication with pencil and paper. For example, suppose 3247 is to be multiplied by 98. In the older desktop machines, this multiplication required two stages of operation. In the first, 3247 was entered into an accumulator (or register) eight times. In the second, the accumulator was shifted one place to the left and then 3247 was entered into the accumulator nine more times, for a total of seventeen additions. In order to speed up the process, James Wares Bryce of IBM developed a clever way of reducing the number of machine cycles, introducing the method of partial products and making use of IBM's new multiple-contact relays. In essence, the IBM engineers "endowed the machine with a multiplication table and were able to reduce the number of separate additions to one more than the number of digits in the multiplicand." Thus, in the case of 3247 × 98, the number of operations was reduced from seventeen to five. One operation is needed for each of the four digits in the multiplicand (3, 2, 4, 7); then there is the final addition of four partial products. A stored multiplication table first yields the product of each digit in the multiplicand multiplied by 8 and then repeats the process for 9. The key to the method is that these individual products are entered in a special way:

$7 \times 8 = 56$  $\qquad$  $7 \times 9 = 63$
$4 \times 8 = 32$  $\qquad$  $4 \times 9 = 36$
$2 \times 8 = 16$  $\qquad$  $2 \times 9 = 18$
$3 \times 8 = 24$  $\qquad$  $3 \times 9 = 27$

The process of multiplication consisted of forming four "partial products," placing them in certain positions, and then adding them:

| | | | | | | |
|---|---|---|---|---|---|---|
| $L_1$ | | 2 | 1 | 3 | 5 | |
| $R_1$ | | | 4 | 6 | 2 | 6 |
| $L_2$ | 2 | 1 | 3 | 6 | | |
| $R_2$ | | 7 | 8 | 6 | 3 | |
| | 3 | 1 | 8 | 2 | 0 | 6 |

The numbers 2 1 3 5 in the $L_1$ line are the left digits of the products in the left column above (reading from bottom to top), and the numbers 4 6 2 6 in the $R_1$ line are the right digits of the same numbers. Therefore, the diagonals in these two lines, reading from upper to lower lines, are the products 24, 16, 32, and 56 of the left column. The numbers 2 1 3 6 in the $L_2$ line are the left digits of the right column, and the numbers 7 8 6 3 in the $R_2$ line are the right digits of the right column. As before, the diagonals in these two lines, reading from upper to lower lines, are the products 27, 18, 36, and 63 of the right column. Adding these four partial products yields the desired result, $3247 \times 98 = 318,206$. Note that in this method of multiplication the problem of carry is postponed until the final addition.

In the 601, IBM's most popular machine of the 1930s, the partial products in lines $L_1$ and $L_2$ were summed (or accumulated) in a "left" register and those in $R_1$ and $R_2$ in a "right" register, so that only a single machine cycle or "addition cycle" was required "for each pair of lines, after which a final addition cycle would add the contents of the left and right registers to produce the final answer." The 601 was an "electric multiplier." Introduced in 1931 and improved in 1933, this machine was originally known as an "automatic cross-footing punch." "Cross-footing" was a bookkeeping method for summing a table of numbers row by row. The 601 could read numbers $p$, $q$, $r$, and $s$ from the several fields of a punched card and produce results of the form $pq$ or $p + q + r$ or $pq + r + s$, and it could subtract by means of complement arithmetic. Multiplying by the method of partial products, its maximum capacity was eight digits in the multiplicand and eight in the multiplier. With an eight-digit multiplicand and an eight-digit multiplier, the 601 could perform 600 multiplications per hour, 10 per minute—in other words, it required 6 seconds per cycle. In the ASCC/Mark I, thanks to Bryce's innovation, multiplications by two digits of the multiplier were performed simultaneously, in a single machine cycle; this cut the machine time needed for multiplying by nearly half.

In a description of the ASCC/Mark I written in 1947, Richard Bloch, who had been the principal programmer of the machine during the war, wrote that the machine could "perform the following basic arithmetical operations: addition, forming the negative of a quantity by reading its complement on 9, forming the positive or negative absolute value of a quantity, subtraction, multiplication, and division." All these operations "except the last two," he noted, were performed by the "relays controlling the storage registers." A "separate unit" was "provided for multiplication and division." Bloch then described the operations for multiplication:

To perform a multiplication, the successive odd and even nonzero digits of the multiplier are sensed in pairs, and partial products are separately accumulated by adding, with appropriate shifting, the corresponding multiple from the stored table. The final product is then formed by addition of the two partial products, and the algebraic sign is automatically prefixed in accordance with the signs of the input quantities.

This description, although written in technical jargon, is almost identical to the above presentation of how multiplication was performed on the IBM 601. With regard to division, Bloch merely noted that it was "performed by selective subtraction in a manner analogous to the longhand method of grade school." The Bryce divider on the ASCC/Mark I was essentially a device for automating the steps of long division, just as Bloch reported, but with an ingenious way of reducing the number of machine cycles.

The IBM business machines of the 1930s and the early 1940s were thus able to perform three of the four elementary processes of arithmetic—addition, subtraction, and multiplication—in the manner just described, but they were not able to perform division in a rapid, practical manner. IBM's official historians record that the IBM multiplying punches, like other business machines of that era, "were not designed to perform division," and that accountants of the day "made use of reciprocals wherever possible."

Thomas J. Watson Jr., perhaps the most important figure in projecting IBM into the computer business, refers in his autobiography, *Father, Son & Co.*, to IBM's problems with dividing. The occasion for his remark is a discussion of his own role in getting vacuum-tube technology introduced into the IBM product line. Watson was especially proud of the part he had played in getting IBM into electronics—notably in the introduction of the first IBM machine to use electron or vacuum tubes, the so-called IBM 603 Electronic Multiplier Punch. Within a year of the introduction of the 603, he wrote, "we got past the gimmick stage." That is, IBM "figured out how to make electronic circuits not only multiply but *divide*—a job that was almost prohibitively expensive to do mechanically."

Similarly, George Daly, an assistant engineer to Clair Lake, and one of the developers (along with James Wares Bryce and others) of the IBM 601, pointed out in an IBM interview that at the time when the ASCC/Mark I was being designed division had long been a conspicuous shortcoming of IBM machines. It was done by a method known as "progressive subtracting," a kind of mechanical analogue of the successive steps in the method of dividing used by humans. This way of dividing, according to Daly, was used rather generally in business machines of the time, including those made by Monroe.

Daly's description of progressive subtracting is succinct and clear: "The operator entered a dividend and divisor and subtracted the dividend from the divisor," starting from the left, "as many times as necessary until" the product would go "below zero." At that point, "the machine reverses its operation and adds the divisor once," whereupon "the net number becomes the first figure of the quotient." This process would be repeated for each digit in the dividend. As Daly remarked, this method "is all right; it gives you accurate results but it's terribly slow." Obviously, if this mode of division were to be incorporated into Aiken's machine, each division would occupy far too many machine cycles and be far too slow.

Around the time that Howard Aiken approached IBM seeking support for his proposed machine, an efficient way of dividing had finally been produced. The brainchild of J. W. Bryce,[1] it was just then being given field trials. The details of its operation were still proprietary, and, as has already been noted, when Aiken was introduced to IBM's product line in January 1938 the question arose of whether he should be allowed to learn about it.

Some sense of the situation at the time when the ASCC/Mark I was being designed can be gained from the IBM interview with George Daly. When Frank Hamilton was assigned to build Aiken's machine, Daly said, he "didn't know anything about multipliers and dividers." He had been "an accounting machine man." At the time, Daly reminded his interviewer, IBM accounting machines "added and subtracted" but "they didn't multiply or divide." IBM had a multiplier which was a separate machine, described by Daly as "a fairly good single purpose machine, but not tied into systems." Hamilton had no experience with these multipliers, Daly said, and so it became Daly's job to instruct Hamilton in their design and use. He showed Hamilton "how we could develop partial products and transfer them together to build up products." Later, this machine process was enlarged so that the eventual Aiken machine could make use of "multiple entries, multiple partial products which were especially useful in dividing."

Aiken suggested a method of performing division by computing reciprocals of the divisor. This process was to be based on a method of iteration, generally known as the Newton-Raphson method, that produces closer and closer approximations. Aiken's proposal was rejected on the ground that Bryce's mode of division would use fewer machine cycles.

Frank Hamilton's 1944 historical memorandum describes Bryce's method succinctly as "based on all nine multiples of the divisor being stored in the machine and chosen dependent on the relation of their

---

1. Throughout this chapter, I refer to IBM's divider (sometimes known as the "plate" divider) as if it had been the brainchild of J. W. Bryce. However, George Daly, in his IBM interview, attributed this invention and the companion device for multiplication to Clair Lake. According to an IBM pamphlet about it, the ASCC used "many basic units . . . invented or developed by IBM engineers," among them a "Multiplying Machine, invented by Bryce in 1934," "Dividing machine, invented by Bryce in 1936," and "Multiplying Dividing Machine, a combination of Mr. Bryce's above mentioned multiplying and dividing machines in a single machine, utilizing the same mechanisms for both functions, invented by Bryce and Dickinson in 1937." There was also a "Ratchet Type Plate Counter, invented by Lake and Pfaff in 1935."

value and the value of the dividend." The "chosen value would then be subtracted from the dividend." Hamilton observes that this "method of dividing is the one that is employed on the machine at the present time." In practice, the ASCC/Mark I produced a table of the first nine multiples of the divisor and then, in a single operation occupying the first cycle, simultaneously compared these with the dividend (or, in later stages, with the remainder). In the second cycle, the highest possible multiple was selected and subtracted from the dividend. If both the divisor and the dividend had 23 nonzero digits and the division was carried to 23 digits, the total time for the division would be 15.3 seconds, corresponding to 51 cycles. Hamilton wrote that the ASCC/Mark I presented "an excellent opportunity to incorporate" the Bryce system into a working machine and to test "the value of the increased speed."

In 1973, Aiken told Tropp and me how he had invented a quite different method of division by machine. During one of his early regular visits to Endicott, he told us, he became aware that the method of dividing by progressive subtraction would occupy so many machine cycles that it would slow the machine's action to an intolerable degree. He recognized that a better system for division was needed if the new calculator was to function efficiently. After he became aware of the problem, he stayed up most of the night in his Endicott hotel room, eventually coming up with a solution. Basically, he decided to use the Newton-Raphson method, the customary resort of applied mathematicians. The Newton-Raphson method consists of making an approximation to the answer and then improving it in successive steps until the result is as close to the answer as one wishes. If $c$ is the approximate root of an equation $f(x) = 0$, the Newton-Raphson method tells you that a better approximation will be $c - [f(c)/f'(c)]$. In order to be sure that Tropp and I understood what he was talking about, Aiken grabbed a pencil and a pad of paper and, with evident joy, solved a problem for us using the Newton-Raphson method.

When Aiken arrived at IBM the morning after his late-night session, he immediately showed Frank Hamilton and Ben Durfee what he had accomplished, explaining how he had applied the Newton-Raphson method in solving the problem of division. He indicated that his proposal would require far fewer machine cycles than the cumbersome method of successive subtraction.

Hamilton and Durfee, not having studied the calculus, had never heard of the Newton-Raphson method. Aiken taught them how to use the method, showing them how it worked in a couple of simple examples. Durfee, amazed, got up and left the room. He did not return for several hours. When he returned, bursting in on Aiken and Hamilton, he could hardly contain his enthusiasm. "It works, Mr. Aiken!" he exclaimed. "It works every time!"[2]

Even though the Bryce divider produced results somewhat more quickly than the older ways using progressive subtraction, it was still very slow. Addition or subtraction with the ASCC/Mark I required one machine cycle (0.3 second). Multiplication required 20 cycles (6 seconds). But division, as we have seen, could require 51 cycles and take 15.3 seconds. Eventually, Aiken removed the Bryce divider.

Aiken discussed the problem of division in the course of the celebrated Moore School lectures, given at the University of Pennsylvania in 1947. He began his second lecture with a presentation of the "Newton-Raphson iterative process," demonstrating how the Newton-Raphson method could be used to produce a table of reciprocals and explaining how such a table could "be constructed with a very small amount of apparatus." He then demonstrated how an equation he had developed could "be used to eliminate the necessity of constructing dividing machinery." He noted that this was "especially important since experience has shown that the relative frequency of multiplication and division is nearly 20 to 1." Accordingly, "a dividing unit represents a relatively inefficient [that is, very little used] piece of apparatus."

In the 1973 interview, Aiken repeated something he had said many years earlier at the Moore School: that "in division the frequency of occurrence is much lower than addition or multiplication." There "is no question of it," he added. IBM "handled this by computing the reciprocals by hand and then punching a reciprocal on a card," finally "multiplying all the arguments by the reciprocal." Another way, he reminded us, was "to look it up in the tables." In fact, "there were some computers and some mechanical calculators that published tables of reciprocals to sell to you right along with the machine."

In introducing this subject, I had mentioned that IBM knew how to add and subtract as well as multiply, but didn't know how to divide. Aiken commented: "You said that they could add and subtract. They couldn't subtract." The ability to subtract wasn't "necessary," he said,

---

2. Aiken seems to have delighted in telling this story again and again. I first heard it from one of my older colleagues while I was a young instructor in physics at Harvard. During the interview, I goaded Aiken into recalling it so as to get it on the record from a primary source.

"because of a great invention by Crompton . . . called 'end around carry.'" (This was a means of adding a 1 to the result of subtraction by complement arithmetic, thus correcting an error that had arisen from the use of complements on 9.) At this point, Tropp asked Aiken whether his "method of multiplication with reciprocals [had been] rejected because the circuit needs were going to be difficult." Not at all, Aiken replied; it "was just rejected out of hand." He then explained that "Bryce [had] devised his dividing machine, which they wanted to put in Mark I." Aiken had "wanted to argue about it," but to no avail. He then reminded us that "later, when we built the relay multiplier, which is much faster, we yanked out the divider and the old multiplier too, for that matter. And that greatly increased the speed and flexibility of Mark I."

In the second of his Moore School lectures, Aiken showed that "even if division and square root units are available" in a machine, "it may yet be more efficient to employ the iterative process in the computation of reciprocals and square roots." He explained how the "iterative method" can be used "directly" to produce the desired results efficiently, so that there is no need to have "a single piece of specialized equipment" built into a calculator for this purpose. Aiken concluded this portion of his lecture by mentioning a recent experience that showed how, even when a computing machine had a special unit for division, the iterative method could be used to increase the accuracy of the result. Recently, he said, Mark I had been used to solve a problem "submitted by the Navy Department." Mark I had then been equipped with a divider. Aiken reminded his audience that Mark I had originally been built to produce tables with "a maximum accuracy of 23 columns." The Navy's problem, however, "required a higher accuracy," which was easily obtained in "the case of addition and subtraction" by simply "ganging two adding counters together to produce an accuracy of 46 columns." The "multiply unit already contained 46 columns in the product counters." Hence, "this unit, together with the ganged adding counters, could be used to obtain 46-column products." The dividing unit was used to find "the reciprocal of the 23 highest-order columns of a 46-digit number"—a result which was then "improved by iteration to yield reciprocals and subsequently quotients of correspondingly high accuracy." This example showed how "the method of reciprocals" was used for "doubling" the "accuracy."

# *11*
## *Construction of the Machine*

Although the first appropriation was made on 12 May 1939, it took a long time for the machine to be designed and even longer for it to be built. There were continual thorny problems of operating technique and a constant battle between an ideal machine and the limitations of cost. Francis Hamilton noted in his historical memo that the IBM engineers eliminated certain features of Aiken's conception in order to minimize costs.

During his interview, Hamilton recalled that Aiken showed up one day with three typewritten sheets of problems he said "would require approximately ten years to complete" with "our production alphabet plate operating at 150 cycles per minute." At the time, Clair Lake had developed his "new ratchet-type counter" to a point such that he and Hamilton decided to use it in the new machine. With this new device, Hamilton told Aiken, the machine would be capable of running at 200 cycles per minute. Aiken was visibly pleased. Doing a little mental arithmetic, he replied that "instead of requiring ten years on the job" the machine could now cut the time down to "seven and one-half years." Among the other IBM innovations Hamilton described were a tape feeding mechanism and tape rack (which he had designed), a value tape mechanism, a punch for preparing the sequence tape (invented by Lake and Hamilton), and a new relay (designed by Lake). This list indicates the degree to which the new machine was constructed of components that were not standard IBM parts and had not as yet been tried and tested in IBM's business machines.

The IBM interview with Hamilton and that with Benjamin Durfee make it possible to reconstruct in some detail just how the great new machine actually came into being. At each stage, Aiken and Hamilton would discuss the problem, Aiken setting forth the "mathematical requirements," the kinds of arithmetical operations needed and their sequencing. Then Hamilton would indicate the hardware available and the way in which the desired sequence of operations could be performed. Lake was almost always present at these meetings and would either approve or disapprove, frequently making suggestions of his own. Durfee said that he did not attend these planning sessions. All of the "planning of the machine," Durfee remembered, "was done through Hamilton and Aiken, with Mr. Lake present probably all of the time to make suggestions and give his agreement or not."[1] The reason Durfee gave for his not attending these planning sessions was that there was no point to his "sitting there all day listening to something which probably I didn't understand." After the meetings, when decisions had been made, "the mechanical design people went to work." Durfee would lay out the circuits and do the actual wiring.

According to Durfee's comments in an interview with Larry Saphire, the "first problem" in getting started was always "to select the hardware." IBM, said Durfee, was then the company "best equipped to build the machine because they had the hardware." But Durfee did indicate that in many cases the standard IBM components were not well suited to the purposes Aiken had in mind. As an example, Durfee mentioned "the accumulator" used by IBM, which was "not very applicable to the Mark I."[2] Lake "had been experimenting with the so-called plate accumulator," and it was proposed to "tool up this plate and use it on the Mark I." This was done as soon as approval was given by management to go ahead with actual construction.

Assuming that the machine would use the Lake plate accumulator and also the "Lake wire contact relay," Durfee and his co-workers made a first estimate of the cost of the new machine. It was soon obvious that management would not find that cost acceptable. Accordingly, Durfee reported, they reduced the size and complexity of Aiken's proposed machine, which was "three times the actual size of the Mark I that was essentially built." Originally, for example, Aiken had proposed that there be "three multiply-divide devices"; the IBM engineers reduced this to one. Aiken had specified more than one sequence control unit, or "sequence tape control unit," but finally only one was used. (This alteration did not limit "the capacity of the machine,"

1. Throughout the interview, Durfee referred to "Hamilton" and "Aiken" without any title but always spoke of "Mr. Lake."
2. Throughout Saphire's interview with Durfee, both men always referred to the machine as "the Mark I."

Durfee said, but it did reduce "the speed in getting the results.") Durfee stressed the concern among IBM engineers that the cost might easily become "prohibitive." Accordingly, their goal was to prevent any "unnecessary risk" that might arise from "building more than we needed."

Durfee himself did not decide "what would go into the machine." This was part of the job of "cost estimating," which was done by Lake and Hamilton in consultation with Aiken. They "finally decided that the basic capacity would be 23 digits, with a 24th digit for algebraic sign control with products of 47 digits including algebraic sign control." (Durfee did not give any reasons for the choice of the number 23. In describing the 72 storage registers, he noted that each one had a "24-column capacity," divided into "two groups of 12 columns each.")

Durfee said that he "worked on all the circuitry from the start." The planning "went along from day to day" in such a way that he and his crew "never had to tackle the whole machine at once," which would have been overwhelming. They "took a little bit at a time." It was decided to speed up the arithmetic of multiplication by using the method of partial products, then IBM's normal practice. "We used right- and left-hand receiving counters," Durfee said, "so that the multiplying process could proceed two digits at a time." Thus, "a 23-digit multiplier, which was the capacity of the machine, . . . required twelve machine cycles plus a gathering together cycle in order to combine the left and right products."

According to Durfee, Aiken didn't have much to do with the actual choice of counters and other devices, nor did he suggest the actual circuitry of the machine. "This was more or less left up to me," Durfee recalled, "because it wasn't a case of discovering too much new in design." The "design of the plate accumulator and other devices in the machine," Durfee reported, implied that the speed be "200 cycles per minute." Accordingly, "whatever you could do within one machine cycle, it required that time."

Durfee made an important distinction between Aiken's contribution (which Durfee designated "the mathematical problem of the machine") and the actual choices of hardware elements and the circuits (all of which were made by the IBM engineers). Durfee was not party to any of the "original discussions" with Aiken. At first, these discussions were between "Bryce and Aiken"; "later on" they involved "Hamilton and Aiken and Mr. Lake." Durfee was quite explicit about the distinction between the theoretical or general level of activity and the practical level.

In Endicott, Lake, Hamilton, and Durfee showed Aiken "the many uses of relay contact points, top counter readouts, and emitters, and the results which might be accomplished by their use." Aiken became "very enthusiastic" and spent many hours drawing "sketches of circuits" to accomplish certain results. Some of these were actually incorporated in the machine design, according to Hamilton, but in many cases the IBM engineers had to revise them "to accomplish exactly the results desired." Aiken kept providing constants and other information during the fall, winter, and spring of 1939–40. He spent all of June, July, and August 1940 at Endicott, during which time the planning for the machine was completed. Aiken's time was spent chiefly in checking the various suggested changes in the machine. During this summer Aiken gave a course in mathematics in the IBM school for employees in Endicott.

Aiken also visited the Endicott laboratory in October and November of 1940. In a letter to Lake dated 15 November, Aiken referred to the new relay designed by Lake, which was going to be tried out on the new machine. He had shown Lake's new relay to some Harvard colleagues, and they had complimented Lake on the invention. He planned to show some faculty members pictures of "the tape mechanism and the multiplying and relay panel." On 21 November, Aiken wrote that he had shown pictures of the tape mechanism and the panel to members of the faculty, who had "duly admired" IBM's achievement. "Every member," he reported, "was enthusiastic over the progress and anticipated the day when the Calculator would be placed in service."

In February 1941, Bryce informed Lake, Hamilton, and Durfee that he intended to bring in a lawyer "to discuss the patentable features of the Calculating Machine." The lawyer, he said, was interested in seven specific features of the new machine.

On 6 March 1941, Aiken sent Hamilton 36 logarithms and 21 sine values to be stored in the machine for use in calculating values of those two functions. He also specified his requirements for interpolation. A month later, Aiken was called up to active duty in the Navy, having enlisted in the Naval Reserve some time earlier. Resplendent in his new uniform, he visited the Endicott laboratory on 24 May, announcing that he would have "very little time to spend on this machine from that time on."

Later in 1941, Aiken designated Robert Campbell to serve as his deputy during the final stages of construction of the machine. Campbell, then in his second year of graduate school at Harvard, had more experience than this description might suggest. After receiving an S.B. degree from the Massachusetts Institute of Technology in 1938, he had gone to Columbia University, where he had received an A.M. degree in physics in 1940. He entered Harvard during the academic year 1941–42 as an advanced graduate student in physics, a teaching fellow in physics, and a tutor in the Department of Physics. He was designated a teaching fellow in electronics in 1942–43 and an instructor in electronics in 1943–44.

Campbell met Aiken for the first time during the Christmas recess of 1941–42, just weeks after Pearl Harbor. While Campbell was in New York, with his wife, one of his Harvard professors got in touch with him and asked him to meet Aiken in Grand Central Station. Campbell had never met Aiken, but he later learned that Aiken had chosen him from a list of graduate students for the job of getting the calculator completed. Aiken was traveling by train from Boston to Yorktown with his fiancée, Agnes Montgomery, and had to change trains in New York. Since there was a wait between trains, Aiken decided to take advantage of the occasion for an interview. Liking Campbell from the start, he asked him to serve as his deputy in the final stages of the completion of an automatic calculator which IBM was constructing for Harvard. Campbell accepted at once.

Aiken was fortunate in getting Campbell to take on this assignment. Trained at MIT, Columbia, and Harvard, Campbell had a solid background in physics and was adept at applied mathematics. Until Aiken was reassigned to Harvard by the Navy (in the spring of 1944), Campbell was in charge of the machine, programming and running the first problems. He also was responsible for working out all the initial errors and reliability problems. His share in the final testing and operation of the ASCC/Mark I was much larger than is apparent in most histories of the computer.

Campbell made his first visit to IBM at Endicott in January of 1942. At about the same time, Professor Chaffee made arrangements with Clair Lake to have Robert Hawkins of the Cruft Laboratory machine shop assigned to work at Endicott with the IBM staff during the final construction and testing of the machine. In this way, Chaffee believed, Hawkins would learn to operate and to maintain the machine. After the machine was delivered to Harvard, Hawkins became a key member of the staff of the Computation Laboratory.³ Hawkins was an unimposing member of the eventual staff of the computer project. Everyone else was in Navy uniform or fully dressed with jacket and necktie, but not Hawkins; he was always to be seen with his shirtsleeves rolled up above the elbows, often smoking a huge corncob pipe.

Durfee recalled that Aiken and "a couple of his assistants" came from Harvard to Endicott and, working with Durfee, "made up program tapes for five test problems." Campbell, however, claimed that he "programmed the five test exercise problems,"⁴ and that Aiken "was not involved."

Despite some initial signs of enthusiasm on the part of IBM, progress on the machine was very slow. Aiken first came to IBM in the autumn of 1937, proposing to Bryce that IBM build his machine. The contract was signed in the spring of 1939, almost two years later. Actual design and construction began in earnest in May 1939, but the machine was not completed until January 1943. Why the long delays? Soon after work was begun, war broke out in Europe. Before long, IBM's production facilities were being converted to manufacture items necessary to the war effort. The war shifted all of IBM's priorities. IBM's wartime projects were of two sorts. One was the production of war materiel, including machine guns for the infantry. The other included developing and producing some specially adapted IBM machines that could serve for calculations directly needed by the armed forces. These included machines for the Manhattan Project, a special rapid calculator for the Aberdeen Proving Ground, and even some "special card reading and punching units" for "an undisclosed project underway at the University of Pennsylvania" (which later proved to be ENIAC). Since the Aiken machine did not seem to serve a pressing wartime need, it received a low priority rating at IBM. Hamilton recalls in his historical memorandum that in September 1942 he was "assigned to war work," but that he maintained "his interest and activities on the Calculating Machine with what time he could spare."

Another—and perhaps a more important—reason why progress was slow, according to IBM's official historians (Bashe et al.), was that the "project was unpopular in the Laboratory." No one was ever going to gain any company recognition in the normal way from the new ma-

---

3. He continued in this important role for about 20 years.

4. One of these problems involved the "mathematics for building the Manhattan Bridge in New York." Another "had to do with one of the planets."

chine, since no product line or revenue was expected to come from it. Everyone at IBM knew that the machine was not going to inaugurate a new product line. It was just a gesture from Mr. Watson toward the scientific community. There was no spur to achievement such as would have existed if it had been thought that work on the machine might lead to improvements of standard IBM machines. It was obvious, therefore, that work on this project was not the sort of activity that would lead immediately to a higher position, a bonus, an increase in salary, or even special corporate recognition. Hence, technicians and engineers weren't striving to work on it. Accordingly, when IBM developed its wartime priorities, the calculator was not among them. Furthermore, there was a real feeling in Endicott that Aiken's machine was far too complicated to function as rapidly and accurately as Aiken had conceived. Hamilton later noted that the project had been considered too "screwball" to be successful.

Yet another reason for the slowdown is that neither Lake nor Hamilton nor Durfee really understood the kind of work for which the machine was being designed. They had no sense that the device they were constructing might be uniquely useful in the war effort. No one at IBM seems to have had a true appreciation of the full significance of the machine they were designing and constructing. They did not envisage the new machine as uniquely qualified to do certain jobs of utmost importance to the war effort. Furthermore, after he had been called to active duty in the Navy, Aiken was no longer in a position to exert pressure on IBM's engineers or executives to complete the machine. Although Campbell kept in more or less continuous contact with Hamilton, he did not have as much clout as Aiken, being only a second-year graduate student at Harvard.

When Campbell took on the new assignment, the basic design of the ASCC/Mark I was nearly complete. What remained was to goad IBM into getting on with construction and final testing, solving various problems as they arose. Campbell recalls regularly traveling by train from Cambridge to Endicott for two years, during which his main jobs were to answer questions, to test new components before they were introduced into the machine, to check the circuit diagrams and then to make sure they were being followed, and to be a general troubleshooter.

It was also Campbell's job to produce certain requirements for the calculator. For example, on 23 March 1942 Campbell transmitted "the values of the powers of 10 needed for computing anti-logarithms." On

11 April he wrote up a series of notes that showed the existence of a fundamental flaw in the design. He showed the IBM engineers "why the machine as originally conceived was wrong for calculating exponentials, logarithms, sines, and interpolation." According to Hamilton, Campbell's "notes also included his recommendations for changes"—changes that were duly "incorporated in the machine."

For Hamilton and Campbell, it wasn't always easy to get responses from Aiken by mail. According to Hamilton's memorandum on the history of the calculator,

> During the time in which Mr. Aiken was in the Navy several questions pertaining to the mathematics of the machine came up on which we desired his advice, but it was impossible to get any reply to correspondence which we sent him. Mr. Hamilton even went so far as to ascertain from the Bureau of Navigation, US Navy, if Lieutenant Commander Aiken had been transferred, and, if so, if we might have his new address. He had not been transferred, and Mr. Hamilton was so informed by the Navy Department.

Previously, Hamilton had reported that the machine was complete and was fully operational by the end of 1942. However, it was on 1 January 1943 that the machine solved its first real problem: to compute and graph "the time required to build up the current in an inductive circuit based on an equation which required the machine to multiply, divide, add, subtract, compute logarithms and antilogarithms."

Hamilton's jubilation at the success of the above-mentioned trial was dampened by the lack of a response from Aiken. Both the "graph and the continuous values computed to construct the graph" had been "transmitted to Lt. Commander Aiken," and "no acknowledgment of receipt of this information was ever received." The equation in question, "not one that might require a practical application of the machine," was "an expression of enthusiasm on Mr. Hamilton's part that the machine had sufficiently progressed so that it was able to compute this equation."

In the opening paragraphs of a letter written on 12 October 1943 by Harlow Shapley to George Henry Chase, a Harvard professor of archaeology who was then Dean of the Faculty,[5] Shapley remarked that "for a year or so" he had "been entirely out of the picture with respect to the interrelationship of Harvard University and the International

---

5. Not to be confused with the George Chase mentioned above.

Business Machines Corporation." Noting that the "work on the big calculator has continued," he mentioned in particular that "the men in the physics department" were "directly associated in the tests and the constructional details." He explained further that "indirectly" he had "occasionally heard of the progress." He was aware, for example, that Theodore Brown still served as "one of the connections" between IBM and Harvard, and that Robert Campbell had made trips to Endicott "in an official capacity" and knew "the details." At the same time, Shapley admitted that President Conant "has been terribly busy" and "so have all of us." Shapley was concerned about aspects of the state of relations between the two institutions: "Mr. Edward M. Douglas of the I.B.M. (a vice president, I believe) dropped in at the Harvard Club in New York City. He talked about the machine, and I am afraid about Harvard's *official* attitude toward the machine, with Howard Reid who is a member of the Visiting Committee of the Physics Dept."

Shapley's letter leaves no doubt that there were serious problems. There is also a mysterious hiatus in chronology, evident from Hamilton's historical summary. The new machine, as we have seen, was completed in the North Street Laboratory at Endicott, and it ran its first test problem in January 1943. But not until December 1943 was the machine demonstrated to members of the Harvard faculty. The reason for this delay is shrouded in mystery.

Perhaps Shapley's letter is related to the long delay in turning the machine over to Harvard. There is apparently no record that any real use was made of the machine from January to December of 1943. Hamilton reported only that during the "last half of 1943" the machine was "demonstrated many times to Customer Administrative Schools." This statement implies that IBM officials displayed the new machine as an example of the technical marvels IBM engineers could produce and did not recognize that the machine could be of service in the war effort.

One possible reason for IBM's not sending the completed machine to Harvard at once is that IBM was then devoting all its manpower to war-related production. Transferring the machine to Harvard would have taken a large number of man-hours away from that work. Another factor that might have held up the transfer was a legal one: actual ownership of the finished machine had not yet been officially transferred to Harvard. There was also the problem of securing IBM's patent rights in regard to any patentable innovations.

It is possible to reconstruct some of the problems that beset the relations between Harvard and IBM by sampling the correspondence. On 4 October 1943, Thomas J. Watson wrote a long letter inviting James B. Conant to come to Endicott to inspect the giant new calculator later that month. Watson used the letter to explain something of his own interest in promoting education. He began, however, in the manner of a manufacturer who has completed a product for a customer rather than in the style more customary in educational circles:

> The calculating machine which we have constructed for the University is now ready for delivery, if it meets your requirements. In order that we may make sure that the machine measures up to your specifications we would like you to come to our plant at Endicott, New York, accompanied by members of your staff who are interested in this machine, for a complete demonstration of it.

Then Watson invited Conant to visit IBM for two days of ceremonies and seminars, which appear to have been unrelated to the completion and demonstration of the calculator. He appears to have had no idea that Conant was busy with war work. Watson expressed the hope that on 18 October Conant could be among the "outstanding educators" usually present at occasions such as the exercises that would take place that day—a kind of commencement for young women and Customer Engineers who were to be graduated from the IBM School. He also wished that on 19 October Conant might attend the sessions of an "educational research meeting." Watson then went on to explain to Conant how "an educational system in industry" should supplement the work of colleges and universities by teaching young people how "to perform specific tasks in business." Finally, Watson returned to the subject of the calculator, asking Conant to suggest an alternative date if 18 October was not convenient and giving two distinct reasons for the visit's significance: ". . . we feel it is important for our engineers who have worked on this machine to meet personally with you and your associates. Your constructive suggestions are important in any further development and improvement of this calculator."[6]

On 6 October a copy of Watson's letter was sent to Harlow Shapley by Edward Douglas of IBM "at the suggestion of Mr. Howard Reid, and also because of [Shapley's] interest in the subject." On 12 October, as a consequence of that communication, Shapley sent Dean Chase a

---

6. This seems to indicate that in Watson's opinion the machine had not been fully completed and was not ready for delivery.

brief account of the relations between Harvard and IBM. After a two-paragraph summary, Shapley explained why he was writing:

All of that is prefatory to the statement that Mr. Douglas has sent me a copy of the letter written on October 4 by President Thomas Watson to President Conant, specifically inviting President Conant to come to Endicott to witness the magnificent machine, which is completed and ready for delivery.

Shapley then explained what he saw as the cause of tension between Harvard and IBM:

Mr. Watson doesn't understand the machine, personally of course, and he doesn't expect Mr. Conant or you to understand the machine. But Watson and the engineers who built the machine for Harvard do crave the courtesy (and publicity) of Harvard's official acknowledgment of their existence and their efforts. This is a private letter, of a sort, but its object is to say that I hope something can be done in an official way—the bigger, the better.

At this point, Shapley stressed the importance of Harvard making a real gesture of acknowledgement and appreciation of IBM's enormous labor in constructing the calculator. "It does not suffice to send only a professor of physics," Shapley insisted, "or a remotely related astronomer." In Shapley's judgment, either "Conant or you or [William] Claflin or John Baker" is "indicated, with a supplement of the Head of the Physics Department." Claflin was a well known figure in the world of business and finance, Treasurer of Harvard and a member of the Harvard Corporation, Harvard's primary governing board, and would thus have been a proper emissary. John Baker was dean of the Harvard Graduate School of Business Administration and was also someone well known in business circles; his presence would be a sign of Harvard's recognition of IBM's achievement. Shapley concluded:

Please do not trouble to answer this letter. It merely records that I am a bit worried about this business—disturbed that we may, through being casual, miss an opportunity to annex some very useful, even though somewhat astonishing, collaborators and patrons.

On the carbon copy of the letter, a note in longhand adds: "This letter is anticipated by news from Chaffee concerning the *official* December plans." Obviously Conant, unable or unwilling to go in October, arranged with Watson to schedule a visit for December.

Shapley's letter provides an important key to an understanding of later events. As Shapley shrewdly pointed out, Watson was eager for some recompense for the considerable effort of constructing the machine. IBM had made an enormous expenditure of time and money on a project for which there was no foreseeable cash return. Watson wanted two kinds of reimbursement of a wholly different sort: first, the favorable publicity that IBM would receive for producing this new instrument for science and for its generosity in making a gift of it to the world of academe; second, an expression of true gratitude on the part of Harvard, through high-level participation in some of IBM's internal education activities and a personal visit by the president and other officials (including the chairman of the Physics Department) to express their admiration and appreciation to Watson and to the engineers who had labored to construct the great calculator. Above all, Watson quite properly wanted a gesture by Harvard that would indicate that IBM's contribution was being fully valued and duly appreciated. In the event, the desired delegation did go to Endicott to inspect the machine, and a great dedication ceremony was planned.

One aspect of the relationship between Harvard and IBM that may strike some readers as odd is the minimal participation of Harvard's President James B. Conant in the negotiations and the rather brusque way in which Conant rebuffed Watson's proposal that he participate in IBM's program of education.

There was a kind of cavalier attitude that it was a privilege for IBM to make a gift to Harvard, the country's oldest institution of higher learning, which had recently celebrated the 300th anniversary of its founding. Perhaps of far greater importance, Conant was no longer taking an active part in the affairs of the university, and he had delegated responsibility for the calculator to Harlow Shapley and Dean Harald Westergaard of the School of Engineering. The burden of Conant's responsibilities in the war effort was heavy. He was second in command of America's scientific work, and he was directly responsible for the Manhattan Project.

However, Conant's attitude toward the ASCC/Mark I is still somewhat puzzling. How can he have failed to appreciate that the machine could be of major service in many wartime research and development projects that were in desperate need of computer power, such as the implosion calculations for Los Alamos? How could he have allowed the great machine to stand idle at Endicott for a whole year when there were so many pressing war-related problems to be solved?

Many years after the events, I spent a couple of hours with Conant when he was in Boston for a meeting of the American Association for the Advancement of Science. I asked him directly about his recollections of the problems between Harvard and IBM and about his own role in relation to the use of the new machine. To my great surprise, Conant attributed the difficult relations between Harvard and IBM to disagreement between T. J. Watson and the Dean of Engineering. He had no recollection of the importance of the machine or of its potential use in wartime research.

Of course, by the time of the above-mentioned conversation, Conant had participated in many important experiences that may have either obliterated or diminished his memory of the giant calculator, including the presidency of Harvard during its great postwar expansion, the problems of the peacetime development and use of atomic energy, service as US High Commissioner and later Ambassador to the Federal Republic of Germany, and a monumental study of secondary education in the United States. One can easily understand that in such an eventful life certain memories become dim or completely obliterated. It is also possible that Conant simply didn't want to discuss the episode.

But I believe there is another reason for Conant's apparent amnesia—a deep-seated one. In the 1940s, very few people, even scientists such as Conant, had any real appreciation of the potentialities of computer power. I believe that Conant simply did not have any sense of the importance of the machine that IBM had built and donated to Harvard. There is some evidence to support this statement in the fact that Conant's autobiography does not mention the existence of the giant machine and does not refer to Harvard's pioneering position in the development of the computer. As we have seen, Conant once told Aiken that if he wanted to gain a tenured position at Harvard he had better concentrate on electron physics and not mechanical calculation. Conant, it must be remembered, was a distinguished organic chemist, but his research had never involved extensive calculations.

Although Harry Mimno made annual reports to Conant on the progress of Aiken's negotiations with industry, he did not explain to Conant the potentialities of the new machine for solving pressing problems of science and engineering. Conant, furthermore, did not have a direct role of any significance in the relations with IBM. Watson apparently expected that there would be some direct contact between the president of IBM and the president of Harvard as equals, and that he and Conant would share ideals for advancing higher education. I doubt that he realized the extent of Conant's wartime responsibilities.

The extent of Conant's ignorance concerning Harvard's relations with IBM is evident in the fact that, as Harlow Shapley reported to James Baker, Conant turned to him during the dedication of the ASCC/Mark I to ask who actually owned the machine that they were dedicating.

Basically, however, Conant's lack of appreciation of the ASCC/Mark I and of the significance of its dedication may be taken as a sign of the general failure to recognize the potentialities of machines to supplant human computers.

# 12
## Installing the ASCC/Mark I in Cambridge and Transferring It to the Navy

In December of 1943, Harvard finally acted in response to Watson's letter of the previous October, and a group headed by J. B. Conant went to Endicott to see the new machine and to observe its mode of operation. This visit occurred just about a year after the new machine performed its first calculations. Watson, ill, was not on hand to greet the Harvard emissaries; they were greeted by F. W. Nichol, Watson's administrative assistant, who gave a brief introductory talk. Then Clair Lake demonstrated the machine. According to Conant's letter of thanks to Watson, dated 11 December 1943, Aiken had been present and had given a "discussion of the long years of planning and the vision of your company in cooperating in the development of machines." E. L. Chaffee had told the IBM people of the "careful arrangements for housing the machine in Cambridge" and had discussed "some of the war problems on which the machine will be used."

Conant's letter concluded with a few words about the plans for a ceremony of public unveiling. Harvard, he wrote, planned "to invite a group of distinguished scientists to be present." Conant hoped that Watson would attend so that the university might "thank [him] personally for this great contribution to science." Evidently, Conant had heeded Shapley's advice and was preparing a grand event to recognize Watson's and IBM's gift to Harvard.

Watson replied on 20 December, shortly before IBM began to disassemble the machine and load it onto trucks bound for Harvard. He informed Conant that Norman Bel Geddes & Associates, a well-known industrial design firm, had been engaged to "assist us in designing a suitable covering for the machine." He was agreeable to having the machine reassembled at Harvard "without the case," so that it could be put into service more quickly, but he urged "that no public showing be made of the machine until it is encased." Accordingly, although the machine was actually set up and running by March 1944, the official dedication was put off until the Art Deco-style sheath could be fabricated and put in place. In stark capitals, Bel Geddes's sheath carried the official name: IBM AUTOMATIC SEQUENCE CONTROLLED CALCULATOR.[1]

At this time, Watson evidently envisaged that the ASCC might mark the beginning of a series of other cooperative ventures with Harvard or with members of its faculty. Thus, in his letter of 20 December 1943, Watson told Conant: "Our engineers have another machine in mind which may be of special use to you. I will talk to you about this at a later date."

Late in 1942 or early in 1943, Watson began a search for a specialist in the new field of electronics to be added to IBM's research and development staff. In October 1942, James Bryce sent a letter of inquiry to Harvard Professor George W. Pierce, a great figure in electronics and a successful practical inventor, to find out whether Pierce would be "interested in possibly doing part-time consultation work with the International Business Corporation in the development of electronic equipment applicable to machines manufactured by [IBM]." In declining this offer, Pierce expressed regret that he himself did not "have the necessary time or the proper qualifications to undertake employment in this enterprise." Nor could he think of another qualified person who was not "engaged in active war work." Watson was prepared to pay handsomely for an electronics expert, especially an "outstanding" one. He ordered his staff to find such an expert and to pay him handsomely. "If he gets $10,000," Watson declared, "we would be willing to pay him as much as $25,000." Watson would even "let him keep in touch with his college, going back once a month to give lectures, or, perhaps as often as twice a month." But Harvard did not help him procure an electronics expert.

Harvard's Physics Department found a home for the calculator deep in the basement of the Research Laboratory of Physics. The designated site was known as the Battery Room because it then housed a giant wet-cell battery, which would have to be disassembled.[2]

1. After the war, when the machine was moved to the new Computation Laboratory, Aiken gave it the name he preferred: Mark I.
2. The Battery Room is often said, erroneously, to have been in the basement of the Cruft Laboratory. Actually, this room was in the building known as the Research Laboratory of Physics. More commonly called the Physics Research Laboratory, this building is joined at one end to the old Jefferson Laboratory

When the machine was complete and ready for shipment to Harvard, I was a very junior member of the staff of the Physics Department (an instructor). As such, I participated in the faculty vote to turn over the Battery Room to Aiken and IBM. At the end of the war, the battery was supposed to be reassembled and the room restored to its original function—something that, to the regret of some in the department, was never done. This vote was my first formal contact with Howard Aiken and his machine.

I well remember the excitement that rippled through the Physics Department on 1 February 1944 as the trucks arrived and the parts of the huge machine, many of them in large crates, were transported to the Battery Room. The Cruft Lab's photographer, Paul Donaldson, was on hand to take still photographs and to capture the entrance of the crates on 16-millimeter movie film with his newly acquired Kodak Cine Special.

During the next few months, the machine was reassembled and put into operation. The IBM engineers were on hand to supervise the work. After they left, Ben Durfee stayed on, assuming responsibility for the final adjustments and corrections of the circuits. Robert Hawkins, who had received extensive training at Endicott before the machine was transferred to Harvard, also worked on the final assembling and adjusting.

In this early period, the operating staff of the machine consisted of Hawkins, Robert Campbell, and David Wheatland (a Research Associate in the Physics Department, an amateur electronics expert, and a longtime friend of E. L. Chaffee).[3] In the first stages of the machine's

---

of Physics and at the other to the Cruft Laboratory. The floors of all three buildings are on the same level, so that it is possible to go from the Research Lab to Cruft without being aware of exactly where one building ends and the other begins. The battery room, the site of the calculator, is some distance from the point of juncture, however, so there should have been no doubt concerning the building in question. The confusion may have arisen from the fact that, both before and after the war, Aiken was associated with the Cruft Laboratory group concerned with communication engineering. Furthermore, in the early days the computation project was under the administrative jurisdiction of E. L. Chaffee and his deputy Harry Mimno, both members of the Cruft group.

3. Wheatland, deeply impressed by the achievement of the IBM engineers in constructing the machine, decided that IBM was a good stock in which to invest a sizable sum of money he had recently inherited. He was not able to convince his siblings (except for one sister) to go along with him. When IBM

operation, Wheatland was one of the principal programmers; he worked closely with Campbell, who was in charge of operations. Hawkins was the chief "engineer," dealing with day-to-day problems. In the first published volume of the Annals of the Computation Laboratory, Aiken acknowledged the contributions of Hawkins and Wheatland.

The first two problems set for the new machine came from Ronald King of the Cruft group and from James Baker, an astronomer whose specialty was telescope and lens design and who was then a Harvard Junior Fellow. King recalls that the problem he sent to Campbell was to "work out some integrals." By the time the computations were finished, however, the machine and its operation had been turned over to the Navy, and its programs and outputs had immediately become "classified." King, who had no Navy clearance, could not gain access to the work the computer had done for him. However, he told me, he "got a couple of pages of tables with numbers on them and that was all." According to King, these were "very useful." After the war, Aiken "spent quite a bit of time" doing the computing needed for King's research, including (King told me) "four books of tables" of "integrals of sines and cosines and exponential integrals." Baker's problem was related to efforts to design a high-power telephoto lens for the Army Air Force in which corrections and adjustments would be made for such effects as changes in atmospheric pressure with altitude.

Some of the day-to-day difficulties of getting the ASCC/Mark I to run efficiently are evident from a logbook begun by Campbell and then maintained by Aiken and his staff. On Friday, 7 April, Campbell had a "conference with Dr. Baker of the observatory" in which Baker "gave us the constants for his first lens, and the direction cosines of the incident ray." That afternoon, Campbell "coded tape (about 3 hours)." In the evening, he "punched tape (about 1 hour), and ran problem once." On Saturday, Campbell recorded that there was

---

later became one of the leading "blue chip" stocks, Wheatland was able to use his substantial profits to support his purchases of historic scientific instruments and rare science books. He became one of the foremost collectors of early scientific books, and was the principal founder of the Harvard Collection of Historical Scientific Instruments. The curatorship of this collection has been endowed in his memory and is named after him. Wheatland was always proud of having been a first operator of the ASCC/Mark I. Among Wheatland's papers, there is a letter (dated 27 May 1968), in which Thomas J. Watson Jr. wrote that he was "honored to correspond with the first operator of the Mark I."

"trouble with the interpolation process while running Baker's Problem (Ray tracing problem)." The reason: "Interpolation process stopped twice just after tape had positioned." Also, "Tape had positioned wrongly at least once." The cause was traced to "excess glue on interpolation tape" so that "tape jammed in reading mechanism and tore." Campbell punched a "new sequence tape in afternoon, as [he] had left out iteration for square roots on old one." He also "made several mistakes in punching and left out 2 resets."

On Sunday, 9 April, Campbell ran the machine for 2½ hours and corrected "mistakes in sequence tape." On Monday, "Machine made an error in first computation of Baker's Prob. (for first surface only)," but the repeat "run was all right." However, "Typewriter #2 would fail to space (vertically) a good bit of the time." Also, "Card feed #1 failed to read out most of 9's punched in card; though other numbers were read out." Therefore, Campbell used "card feed #2." He made "3 runs on Baker's problem, using 2 different sets of data," for a total "running time" of "about 4 hours." In the evening, Campbell "calculated sin$x$ for $2.5° = x = 90°$ in steps of $2.5°$ for Underwater Sound Lab's Problem." He checked his results by an ancillary calculation. He found the sum to be "in error by $-244 \times 10^{-15}$." The record of the day's work ended with this note: "Set-up and computing time about 2½ hours."

A week later, on Saturday, 15 April, Campbell recorded the following: "Ran King's Second Prob. (Prob. A), calculating values for cosine tape. Machine made an error in about 3rd set of values, but calculation was continued. Ran about 5½ hours." On Monday, 17 April, Campbell ran "test problems." He took note that "Mr. Durfee from IBM was here to help us find 'bugs,'" and that "Ensign Bloch*, who reports directly to Admiral        , was here also to learn about the machine." (Campbell did not know the admiral's name and so left a blank space.[4] The asterisk was keyed to a note by Campbell, reading: "from the naval res. lab.") Campbell then added that Bloch had "expressed interest in the calculation of Bessel Functions." Here is an early record

---

4. The man was Rear Admiral A. H. Van Keuren. During the 1973 interview, Aiken told Tropp and me how he had lured Richard Bloch to Cambridge to join the staff operating the new machine. Aiken and Bryce had gone to the Naval Research Laboratory "to get the backing of Admiral Van Keuren for some calculations which we wanted to perform." The admiral had assigned one of his young men to show Aiken around and to deal with the problem. Aiken wryly remarked: "I expressed my full appreciation for what Van Keuren had done by recruiting Bloch and taking him away."

that the subsequent tables of Bessel functions were produced at the request of the Navy and not simply because Aiken was enamored of such an assignment.

An entry dated 17 April records what is—so far as I know—the first use of the word "bug" in the context of a computer. Campbell's use of quotation marks may indicate that "bug" was not customary in the literature of applied mathematics or mathematical physics or that Campbell was quoting Durfee. In any case, this term is known to have been used by technologists and mechanics long before the advent of computers.[5]

The calculations performed on the new machine gave results to 23 digits. Because of this large number of digits, Baker told me, he ended up with "a vast excess of accuracy." That is, for Baker and his group, "six decimals were enough; seven would be awfully nice but not essential." But "they ran it" to "twenty-three figures."

In the early years, Baker recalled, it was thought necessary to check the accuracy with which the machine was carrying out its instructions very frequently, whereas today we take a machine's performance for granted and are more concerned with the accuracy of the input data and the correctness of the program. In "those days," according to Baker, as a "hang-over from desk calculators," there "were all sorts of check formulas that went with calculations as well as the original formulas, [so as] to be sure that everything was right before going on." Baker and Campbell soon learned that these checks could be "abandoned because the machine did its own internal checking."

By early spring of 1944, the ASCC/Mark I was in full operation under Campbell's direction. Aiken was still teaching at the Naval Mine Warfare School in Yorktown, Virginia, far from any real active participation in the war and in an assigned post that seemed to offer no possibility of escape. In April, however, there was a dramatic change in Aiken's situation. He was transferred to Cambridge, where his new assignment was to take charge of the computer operation. Aiken, now Commander (USNR), was at last in command of his computer.

---

5. Many years later, Grace Hopper alleged that the term "debugging" had come into general use in Aiken's laboratory after a moth got caught in a relay, causing it to cease functioning. The error was corrected by removing the "bug." In fact there is a dead bug, apparently a moth, pasted into the logbook. But Bob Campbell had used "bug" years before in reference to Ben Durfee's correction of faults ("find 'bugs'").

# 13
## Aiken at the Naval Mine Warfare School

In the spring of 1941, Howard Aiken was called to active duty[1] in the Navy and assigned to the Naval Mine Warfare School in Yorktown. Most of the men there were young, newly commissioned ensigns or lieutenants junior grade—recent college graduates with little or no previous course work in physics and with very meager qualifications in mathematics.

There are a number of different points of view concerning Aiken's first Navy assignment and his performance. In a letter to Aiken, one J. B. Neisser describes him as "armed to the teeth with room-length formulas and ivy-covered Harvard theories" and running "smack into a collection of Dixie dumbbells," not one of whom "knew calculus from corn pone." In August 1955, in a letter to one William Boyd Jr., Aiken recalled his distress on finding out that some of his students would need a preparatory course in elementary calculus. From a letter Aiken wrote to one E. S. Giletti on 28 November 1944, we learn that the primary subject on his mind was the great machine being completed and tested at Endicott.

I was able to find out more about Aiken's teaching experience in Yorktown from John Ladd and Thomas Wright,[2] alumni of the program who stayed on as members of the teaching staff.

1. He was not drafted, as has sometimes been alleged. He had joined the US Naval Reserve some time earlier.

2. Ladd, a college classmate and a longtime friend of mine, completed his Ph.D. at Harvard after the war and became a professor of philosophy at Brown University. He is the author of a highly esteemed study of the Navajo ethical system. Wright became an architect; he has practiced in Washington, D.C., for many years. I was put in touch with him by Ladd.

Ladd's story is typical of the time. Upon graduation from Harvard College in 1937, he immediately entered the Harvard Graduate School of Arts and Sciences to study philosophy. At the end of his first year, he took a job as a teaching assistant at the University of Virginia. When he was assigned a very low draft number (19), he became aware that he would shortly be called to military service. Preferring the Navy to the Army, he decided to enlist in the Navy's V-7 program, the purpose of which was to quickly transform college graduates into officers. "At the time I applied," Ladd recalls, "they had a mathematics requirement for the V-7 program." At the recruiting station in New York he was told that he did not have the proper qualifications in mathematics for the officers training program. "There were a number of people who were in my predicament," he continued. "We all enrolled in a summer school course in mathematics at NYU." Ladd had studied mathematical logic as part of his undergraduate studies of philosophy at Harvard, but the Navy required a formal course in mathematics at the college level. What he took at New York University proved to be a crash course in calculus. When the course was finished, Ladd told me, "We all went over to the recruiting agency and signed up. So that's how I got into the Navy." He was immediately sent to the *Prairie State*, a big old ship, moored on the Hudson in New York, that had been turned into a training school for Navy officers. On completion of the course, Ladd was a fully commissioned officer in the Naval Reserve with the rank of ensign. He was at once assigned as a student to the Naval Mine Warfare School at Yorktown.

Wright was graduated from Harvard College in June 1941 and soon enlisted in the Naval Reserve. He had taken a pre-med program at Harvard. Having studied mathematics and physics, and fulfilling the Navy's other qualifications, he was accepted for officer training at once. Like Ladd and many others, he was sent to the training school on the *Prairie State*. After the customary 90 days, he was assigned as a newly commissioned ensign to the Naval Mine Warfare School. A good student with a sound (if somewhat elementary) knowledge of mathematics and physics, he did well in the course and was kept on at the school, spending the next year as a member of Aiken's teaching staff.[3]

Ladd also stayed on at Yorktown after completing the course, but not on Aiken's teaching staff. His assignment was to take students out

---

3. Eventually he was put in charge of all mine laying and mine sweeping in the Mediterranean.

to sea on a minesweeper so as to give them some practical experience to back up their otherwise theoretical training. Like Wright, he eventually came to know Aiken quite well.

As Ladd remembers it, there were two major "professors" in charge of the teaching program at Yorktown: Aiken and someone from MIT. Aiken was in charge of the course on electricity, the primary subject of which was circuit problems; his MIT counterpart ran the course on magnetism. Ladd recalled that "Aiken gave straightforward lectures" with "not much discussion," and that his presentations were "beautifully clear." "It was like a college course lecture," he said. Aiken distributed "handouts about different kinds of circuits and things."

Wright recalls that by the time he joined the teaching staff, some time after Ladd had taken the course, Aiken was more occupied with administrative duties than with instruction. The officer in charge had been promoted to officer status from the ranks and was not college educated. According to Wright, that officer and Aiken "did not get on."

Aiken, according to Ladd, had no friendly give-and-take with the students; he would just "pop in to give the lecture and then pop out again." In this connection, Ladd reminded me, Aiken was "following Navy hierarchy. He was a lieutenant-commander and I was an ensign. Ensigns didn't rub shoulders with commanders." Once Ladd and Wright joined the teaching staff and were promoted to a rank higher than ensign, however, Aiken socialized with them often.

Wright was married to Ladd's sister, and the two brothers-in-law were very close in those days. "My sister and Tommy were very social," Ladd remembers. "They liked parties; so we used to have parties a lot of the time at their little house on the base." Aiken attended these parties, and so Ladd got to know him quite well. Wright recalls that Aiken often gave parties for members of his staff and their wives or friends. He was a good host, generous with drinks.

I asked both Ladd and Wright whether they remembered Aiken ever talking, in class or outside, about the problems of computing or about his machine (under construction in Endicott). Neither of them had any direct recollection of Aiken's saying anything specific about computing, but both recalled being very much aware that he had invented some kind of mathematical machine. As Ladd put it, everybody around the school at Yorktown had heard "by word of mouth" that Aiken had invented some kind of mathematical machine. Aiken never mentioned the machine in his lectures, but "everybody knew that he had invented something complicated. . . . We all thought him kind of nutty about . . . vacuum tubes, circuits, and so on. . . . We thought on this subject he was a *nut*. That is, I guess, most of his students said that."

Wright remembers that there were about fifty students in each course (electricity and magnetism); Ladd believes the numbers to have been somewhat higher. Ladd recalls that Aiken "was a little snobbish about" his assignment: he was "used to teaching a sort of elite students" and here he had only "run-of-the-mill students"—"pretty good people," but not "fabulous."

I have not been able to verify the story, reported in one history of IBM,[4] that Aiken, accompanied by a crew of Navy men, was sent to examine and disarm a German torpedo that had washed up on the New Jersey coast. Since torpedoes almost always exploded on landing, this was a rare opportunity to examine one of the Germans' new secret weapons and to discover the details of its mode of operation. I assume that it would have been up to Aiken whether to perform a superficial examination and then have the torpedo destroyed or to attempt to disarm it and examine its internal mechanism. Aiken is said to have sent his crew a safe distance away while he first examined the torpedo and then proceeded to disarm it. He reported each step so there would be a record of what he had done in case the device exploded. According to the story, Aiken was asked some time later why he had not ordered his crew to retire to a safe distance and shoot at the torpedo so as to cause it to explode. Aiken is said to have replied that Lieutenant Commanders in the Navy were a dime a dozen, but to find an intact, unexploded enemy torpedo was a unique event.

According to another war-related story, of a very different sort, Aiken was once disguised as a French peasant and spirited by submarine into occupied France on a secret mission to examine a bit of German apparatus that was to be available for only a day. During the 1973 interview, Hank Tropp asked Aiken whether or not this episode actually occurred. Aiken did not answer.

I asked Mary Aiken whether she had ever heard either of these two stories from her husband. Although she had never heard the one about the torpedo, she remembered that Aiken had told her something about a visit to France during World War II, and she thought

---

4. William Rodgers, *Think: A Biography of the Watsons and IBM* (Stein and Day, 1969).

he had gone on at least one submarine expedition related to the placement of mines.

In any event, it is known that Aiken was chafing at the bit in his teaching assignment at Yorktown while the great machine was being built. Aware that Robert Campbell was overseeing the final stages of construction, he wanted desperately to be assigned to Cambridge and put in charge of his machine.

But could a full-time naval officer be assigned to a civilian project? Eventually this problem was resolved by having the computer operation run at Harvard as a Navy unit. Harlow Shapley, who had been so important in establishing Aiken's contact with IBM, was again active on Aiken's behalf. His correspondence shows that he maintained contact with Aiken from 1941 (when Aiken was assigned to Yorktown) until 1944 (when he was transferred to Cambridge). Shapley kept Aiken posted about plans for a science center at Harvard, about relations between J. B. Conant and T. J. Watson,[5] and (in 1943) about a possible radio program in which three scientists—Aiken, Norbert Wiener, and John von Neumann—would discuss the new "Giant Automatic Calculators."

While Shapley was using his influence in an attempt to get Aiken transferred to the computation project, Aiken himself was trying to convince his superior officers of the value of his machine to the war effort. The ultimate success of all these efforts is evident: the Navy took over the operation of the calculator, staffed it with Navy personnel and some civilian employees, and paid all the operating costs, including a rental fee to Harvard.

In his speech at the unveiling of the machine, in August 1944, Aiken paid tribute to "Captain Solberg and Commander Ferrier, who, hearing of the machine's completion and delivery to Harvard, at once took action in the Navy Department to see that the machine was put to work on the solution of problems in the interest of the national defense." It was "for that reason," he added, "that I have been sent back to the University to carry on with the machine." In 1959, in a talk at the Navy's David Taymor Model Basin, Aiken said this: "The Navy's first computer project was the operation of Mark I at Harvard as an activity of the Bureau of Ships." This project "was first recommended by Commander David Ferrier, who was then Officer-in-Charge of the

---

5. Despite an earlier contretemps, Shapley assured Aiken that "all seems to be handled pretty well."

Naval Research Laboratory at Harvard University." Ferrier's "recommendation was made to Admiral A. T. Solberg who approved the program and subsequently supported its development."

In 1973 Aiken told Tropp and me that in the spring of 1944 he had been exceedingly eager to return to Cambridge and take command of the calculator, which now had been tested and put into preliminary operation. He told us that he continually stressed to his superiors that he would be more useful to the Navy if he were in Cambridge in charge of the machine than in if he remained at Yorktown running the Mine Warfare School. One day, he said, he was summoned by a high-ranking Navy officer who asked him why he was not running his own calculator. "I have orders," Aiken replied. Within hours the orders had changed. On 27 April, Aiken sent a telegram from Yorktown to Campbell:

HAVE THE PAPERS
SEE YOU SOON.

In another telegram, he told Shapley

THE MILLS HAVE GROUND EXCEEDING FINE
SEE YOU SOON.

On Friday, 5 May, Campbell wrote in the logbook "Commander Aiken arrived from Washington." Aiken became the first naval officer in history to be the commander of a computer.

# 14
## *The Dedication*

On 17 April 1944, James Conant reported to Thomas Watson that "the calculating machine" had been "put into operative condition." Expressing his appreciation for "the speed with which the machine has been installed" and noting that it "is already being used for special problems in connection with the war effort," Conant nevertheless regretted that the sheath was "still to be completed," and that therefore the "public announcement of the machine and its purposes" would have to be delayed. Watson replied on 21 April that he had "given orders to rush the completion of the cabinet for the machine." He also expressed the "very great pleasure and inspiration" it had been for his "organization to cooperate with you and your associates in connection with this machine." He was "looking forward to being present at the dedication."

Present-day readers may find it odd that the unveiling of a new calculating machine should be called a dedication. This is, however, the name that has been generally used for the ceremony at which Thomas J. Watson of IBM officially presented the ASCC/Mark I to Harvard. This expression does not occur in the Harvard news release, nor does it appear in the official invitations, which referred to "the formal acceptance celebration ceremonies of the AUTOMATIC SEQUENCE CONTROLLED CALCULATOR constructed and developed by the International Business Machines Corporation with the Coöperation of Harvard University." However, "dedication" does appear in letters exchanged by Conant and Watson during the planning of the event—for example, as has already been mentioned, Watson informed Conant on 21 April that he was "looking forward to being present at the dedication."

There is good reason to suppose that Watson was responsible for calling the event a dedication. He apparently liked the idea that he was dedicating the new calculator to the advancement of knowledge by donating it to a great university. A few years later, in June of 1948, Watson used the term "dedication" when another machine, the SSEC, was formally unveiled in the presence of a group of invited mathematicians and scientists. Watson formally dedicated the SSEC to "the use of science throughout the world," and the official record of the event was headlined "IBM's Selective Sequence Electronic Calculator is dedicated to the Aid of Science."

On 24 July 1944, Conant wrote telling Watson of the plans to celebrate the unveiling. On Monday, 7 August, after a "small luncheon to be held at twelve-thirty in the Naumburg Room of the Fogg Museum of Art," ceremonies would take place in University Hall. Conant hoped that Watson would "say a few words." Other invitations that were issued show that there was to be a less select, less formal luncheon at 1 P.M. in the Faculty Club. The formal acceptance ceremonies were scheduled for 2:30 in the Faculty Room of University Hall. Then the guests would proceed to the Computation Laboratory in the Research Laboratory of Physics to see the calculator. Finally, they were to be offered tea in the Faculty Room of the Department of Physics.

The list of invitees to the ceremony included the Navy officers assigned to the machine, some civilians working for the Navy; Harvard's president, members of the governing boards, and administrative officers; the IBM engineers responsible for the machine's construction; other Navy officers (at least three admirals); and the governor of Massachusetts, Leverett Saltonstall.

The Harvard News Office, with Howard Aiken in close consultation, prepared a news release.[1] The release, headed "World's greatest mathematical calculator," bore this statement: "The NAVY, which has sole use of the machine, has approved this story and set this release date."[2] The first four paragraphs stated that the machine would be presented to Harvard by IBM,[3] that it would solve many types of mathematical problems, that it would be presented "by Mr. Thomas J. Watson, president of International Business Machines Corporation,"

1. See appendix A below.

2. Since it was wartime and the operation of the ASCC/Mark I had been assigned to the Navy, Harvard had to get Navy clearance for the event and for any official news releases about the machine. The release date was 7 August 1944.

3. The news office evidently did not consider it necessary to clear the release with IBM.

that it was "new in principle," and that it was an "algebraic superbrain."

"In charge of the activity," the release continued, "is the inventor, Commander Howard H. Aiken, USNR . . . [,] who worked out the theory which made the machine possible." Not only was Aiken designated as "the" inventor; no reason had been given thus far for IBM's being the donor. In fact, nowhere in the news release was it even mentioned that IBM had constructed the machine. Indeed, the mentions of IBM's contribution to the invention were confined to one paragraph:

Two years of research were required to develop the basic theory. Six years of design, construction, and testing were necessary to transform Commander Aiken's original conception into a completed machine. This work was carried on at the Engineering Laboratory of the International Business Machines Corporation at Endicott, New York, under the joint direction of Commander Aiken and Clair D. Lake. They were assisted in the detailed design of the machine by Frank E. Hamilton and Benjamin M. Durfee.

It was mentioned twice that Aiken had been "assisted by Ensign Robert V. D. Campbell, USNR," and that the "code book" for operating the machine had been "written by Commander Aiken, assisted by Ensign Campbell."

During an interview in August 1985, Cuthbert C. Hurd[4] told me that he had gotten a firsthand account of T. J. Watson's arrival in Boston from Frank McCabe, who had been the branch manager of IBM's office on Boylston Street in Boston during and after the war. McCabe, described by Hurd as "a pleasant Irishman," had told him that Watson's wrath had been kindled immediately upon his arrival in Boston on Sunday, 6 August. Watson had traveled from New York by train. He expected that he and his wife would be met at the train by some Harvard dignitaries, or their emissaries, with an official limousine, but when McCabe made inquiries he found that there was no plan for any Harvard official to be on hand. McCabe himself drove down to meet the Watsons in his old two-door Chevrolet. The lack of any formal welcome and the mode of transportation were hardly up to Watson's expectations, and he was noticeably irate. McCabe later recalled that

---

4. Hurd claimed to have been one of the first Ph.D.s to join IBM's permanent staff on a full-time basis (in March 1949). For many years he directed IBM's Department of Applied Science. For biographical information see Eric Weiss, "Eloge: Cuthbert Corwin Hurd (1911–1996)," *IEEE Annals of the History of Computing* 19 (1997), no. 1: 65–73.

it had been raining hard, and he particularly remembered Mrs. Watson's difficulty climbing into the Chevy's back seat.

When Watson was seated in the car, McCabe gave him a paper—either the Harvard news release or one of the Boston newspapers that issued preliminary Monday editions on Sunday afternoon. McCabe recalled vividly that when Watson saw the paper "he exploded." He was enraged to learn that Harvard had presented Howard Aiken as the inventor of the new machine, without full credit to IBM.

It is not possible to determine what paper McCabe gave Watson on his arrival in Boston. It really doesn't matter, though; the Harvard news release and the stories in the newspapers would have been equally infuriating to him. The morning newspapers then published in Boston were the *Globe*, the *Herald*, the *Daily Record*, and the *Post*. All three had derived their stories from the news release, giving a thumbnail sketch of Aiken ("the inventor") and repeating the statements about two years of research for the basic theory and six years of "design, construction, and testing" at IBM's laboratories "under the joint direction of Commander Aiken and Clair D. Lake." The story in the *Post* was the worst from an IBM point of view. Page 1 carried the headline AUTOMATIC BRAIN FOR HARVARD and the subhead "Navy Man Inventor of World's Greatest Calculator." A picture of the machine showed the "inventor, Commander Howard Aiken, USNR, explaining it to Lieutenant Grace Hopper, USNR." IBM appears in this story only in a sentence indicating that the new machine "will be presented to Harvard . . . by the International Business Machines Corporation." There is no direct statement that the machine had been constructed by IBM and no indication of IBM's role in designing or inventing it. Two of the five paragraphs were devoted to Howard Aiken.

Aside from the above-mentioned dailies, the Boston-based *Christian Science Monitor*, which claimed to have based its story on "an interview [with Aiken] prior to the presentation ceremonies," reported that IBM had "built the machine from an idea formulated by Commander Howard H. Aiken, USNR, when he was a self-styled 'lazy' graduate student at Harvard eight years ago."[5] In New York, the *Times* ran a headline about an "algebra machine" and the *Herald Tribune* used the term "super-brain"; both were taken directly from the news release. The *Times* also ran an editorial, titled "Aiken's Calculator," in which

---

5. The *Monitor*'s account was notably unfair to IBM; see chapter 15 below.

the machine was likened to "the perfect scholar in a fascist state, answering questions but incapable of asking them." The *Detroit News*, under the headline "Lazy Ex-Professor Builds World's Fastest Calculator," noted that Aiken had "presented the world's greatest mathematical calculator to the university" and went on to state that "Commander Howard H. Aiken, of Indianapolis, made the presentation to President James B. Conant of Harvard in the presence of high naval officials." Aiken is quoted as saying "If I hadn't bitten off more than I could chew when I picked out a problem for my master's degree, I probably never would have thought of the calculator." Being "essentially lazy," Aiken "figured it would be easier to work out a solution on a machine than on paper." Since he "couldn't find any machine capable of handling the problem," he "simply had to invent one." Apart from a number of gross errors of fact (such as "a problem for my master's degree"), this account is notable for the omission of any reference to IBM. The clipping in the Aiken files is annotated, in an unidentified hand: "If I were Mr. W., I would be a bit mad, too." *Time*, too, failed to mention IBM in its article.

Watson would have had several good reasons for anger at the reports. First, he must have felt slighted that a news release had been issued without any input from IBM. Even more important, it is noted in *IBM's Early Computers*, was "the implied insult to his men in the singular use of the term inventor." Moreover, Watson was prepared to surprise Harvard with a check in the amount of $100,000 to defray the costs of operating the machine. In its time this was an enormous sum—it would have paid at least the salaries of ten of Harvard's highest-paid full professors for at least a year.

When Watson reached his hotel in Boston (the Copley Plaza), he telephoned his Harvard hosts and threatened to boycott the ceremonies. Conant and Aiken, and possibly Harlow Shapley, thereupon rushed over from Cambridge, accompanied by Walter S. Lemmon of IBM, in the hope of placating Watson. Watson launched a furious tirade against Aiken and Harvard. Evidently the representatives of Harvard succeeded in calming him, however. He attended the dedication, giving a star performance. On the morning of the ceremony he visited the installation and had himself photographed with the three IBM engineers who had constructed the machine and with several of the enlisted personnel the Navy had assigned to the project.

Further indications of what happened on this occasion and during its aftermath can be gleaned from correspondence between Harlow Shapley and Walter Lemmon. Lemmon evidently was present on the evening of 6 August 1944, when Aiken and Conant met with the enraged Watson at his hotel in Boston.[6]

An engineer, an inventor, a businessman, and a philanthropist, Lemmon was a close associate of both Shapley and Watson. He was born in New York City in 1896, was graduated from Stuyvesant Technical High School, and in 1917 received a bachelor's degree in electrical engineering from Columbia University. Throughout his career his special interest was radio. He invented numerous devices, including the single-dial tuning control system, a number of radio typewriters, and a forerunner of the short-wave antenna. Like Aiken (who was four years his junior), Lemmon was commissioned in the US Navy—but he served in World War I. Between 1917 and 1919 he was involved in radio and signal training courses and in the development of radio telephone systems. He served as special radio officer for President Woodrow Wilson during the peace conference of 1919 aboard the USS *George Washington*. That experience made a deep impression upon him, convincing him of the importance of radio as a means of promoting education and peace around the world. His establishment of the World Wide Broadcasting Foundation and of the international radio station WRUL in Boston was an important aspect of his effort toward this goal and was the source of his long and close association with Harlow Shapley, whom he first sought out in April of 1933 in order to discuss the question of educational radio. Between 1933 and 1945 he was the general manager of IBM's Radiotype Division.

On 27 July 1944, nine days before the dedication of the ASCC/Mark I, Lemmon wrote Shapley a letter that suggests much about the personalities and the issues involved in relations between IBM and Harvard:

Dear Harlow
Since writing a letter to you last night I had a chance to talk to Mr. Watson about several things including the forthcoming ceremonies at Harvard. I thought you would be relieved to know that he was very pleased with the way the situation was being handled and particularly with the setup for the luncheon and the invitations.

---

6. Another member of Watson's traveling party was Vincent Learson, later to become president of IBM.

He asked me if I could arrange to go up the same day as I think he would like to have me on hand during the ceremony. Inasmuch as the luncheon is such a very small and private affair, perhaps I would be simply "impedimenta" during that part of the day, but I would appreciate it if you could arrange to have the University send me an invitation for the ceremonies. If this should be done through the IBM office, let me know but I recognize that time is rather short and I wanted to cut down on the red tape.

This means that I will probably be coming up with Mr. Watson on Sunday and then will stay over after the ceremonies until Tuesday. I hope we can get together either Monday night or Tuesday as I am very anxious to pursue the preliminary steps on the brochures for our World Radio University.

This letter was answered, in Shapley's absence, on 29 July (a Saturday), by Miss Arville D. Walker:

Dr. Shapley left for Sharon before the morning mail arrived, in which were two letters from you. As he probably will not return until late Monday afternoon I have called up the President's office in regard to an invitation to you for the ceremonies on August 7. The secretary says that your name is on the list and if an invitation did not go out yesterday (as she believed) one will go today.

On 2 August, the following Wednesday, Shapley sent his own "hasty response": "I expect to see you next Monday here in Cambridge at the dedication."

On 4 August, Miss Walker typed a memorandum to Shapley:

Mr. Lemmon just called up. He is arranging a schedule for your New York trip. . . . He is coming to Boston Sunday and will see you on Monday. He would like you to hold a little time for him after the afternoon ceremonies. There are some things he wants to talk about with you, among them something very important that is coming up in Washington concerning short waves. . . .

At 5:30 P.M. the same day, Miss Walker appended to this memo a note saying that a telegram had just been received by telephone from New York asking Shapley whether he could "hold luncheon open for Tuesday, August 8, in Boston."

There is evidence that Lemmon was present that evening when Watson upbraided Aiken, and that he heard Aiken's address at the dedication ceremony on 7 August. His presence in Watson's hotel suite and at the dedication is evident from remarks he made to Shapley on 21 September in New York, which Shapley quoted to Aiken in a note written the next day. According to Shapley's note to Aiken, Lemmon had said this of Aiken:

*128    Chapter 14*

A man who can take that Sunday night berating and then give the speech of the next day—he's got guts! And then along comes this disgusting window-dressing in THINK. Why didn't they have the courage and sense to print Aiken's Monday address?

Conant, Aiken, and Lemmon were able to calm Watson on Sunday night, and Watson agreed to attend Monday's ceremonies as planned.

Watson's explosion in the hotel and his alleged threat to boycott the ceremonies are both in character. In *The Lengthening Shadow,* the authorized biography by T. G. Belden and M. R. Belden, it is mentioned that Watson would easily "fly off the handle." Watson had learned early on that expressing wrath was an effective way to make a point. We don't know whether Watson's tirade against Conant, Harvard, and Aiken was a real outburst or whether it was calculated.

The ceremonies in old University Hall were attended by many dignitaries, including—in addition to university officials and members of the computation project staff—leaders of science and industry and an impressive array of admirals and other Navy officers. The most complete account was published in the *Harvard Service News* (the wartime version of the Harvard undergraduate newspaper, the *Crimson*) in its issue of 8 August. The report noted the presence of a number of flag officers: "Rear Admiral Edward L. Cochrane, USN, Chief of the Bureau of Ships, A. Van Kuren, USN, Director of the Naval Research Laboratory, J. A. Furer, USN, Co-Ordinator of Research and Development, Wat T. Cluverius, USN (ret.), Chairman of the Board of Production Awards, and Robert A. Theobald, USN, Commandant, First Naval District." Among the other distinguished guests were Governor Saltonstall (also president of Harvard's Board of Overseers) and Captain C. H. J. Keppler ("commander of the Navy personnel in the various training schools at Harvard").

In the main speech,[7] Howard Aiken presented a fair and reasoned account of the development of the machine, giving due credit to the IBM engineers. The tone of this speech was vastly different from that of the news release. Aiken graciously recalled that his "first contact" with IBM had been made with James Wares Bryce, who "for more than 30 years had been an inventor of calculating machine parts." He mentioned that Bryce's inventions involved "counters, multiplying and dividing apparatus, and all the other machines and parts . . . which

---

7. See appendix B below.

have become components of the Automatic Sequence Controlled Calculator that you are to see this afternoon." He acknowledged that Bryce, with his vast experience in the field of calculating machinery, quickly saw the value of a scientific machine such as Aiken had proposed, and that he thereafter "fostered and encouraged this project."

Aiken specifically pointed out that "the multiplying and dividing unit included in the machine was designed by" Bryce. He then mentioned that Bryce had placed "the construction and design of the machine" in the hands of Lake, Hamilton, and Durfee. He stated explicitly that he himself had "set forth requirements of the machine for scientific purposes" but that the IBM staff had been responsible for bringing the machine into being.

Mr. Watson took top honors for the day with his magnificent gift of $100,000.[8]

8. According to an anecdote attributed to Harlow Shapley, James Conant turned to him before the official program got underway and asked him whether Harvard owned the machine. "That's up to you," Shapley replied. When it was Conant's turn to speak, he praised Watson for his support of science and thanked him for "the gift"; Watson, in turn, publicly announced the gift of the machine to Harvard. The astronomer and lens designer James Baker told me he had heard this from Harlow Shapley, who had sat next to Conant during the ceremonies.

# 15
## The Aftermath

Although the dedication ceremonies went off without a hitch, everyone at Harvard was aware that Watson had not been fully appeased. James B. Conant, who had made plans for "an extended trip at the conclusion of the ceremonies," called Harlow Shapley into his office and asked Shapley to represent him in "the further efforts to clear up the feelings and misunderstandings with regard to the miscarried publicity." Conant delegated "the responsibility for Aiken's part in the show" to Harald Malcolm Westergaard, Dean of the Graduate School of Engineering.

Westergaard seems to have been an able administrator, but he was not very wise in the ways of the world. On Saturday, 12 August 1944, five days after the dedication, he wrote to Watson expressing "the deep gratitude of many members of Harvard University for the magnificent gift of the great calculator." He "regret[ted] exceedingly that the news story released for August 7 was inadequate in emphasis on the role of IBM and its men." He explained that the usual "means of communication" of the faculty members of Harvard University were "books, journals, lectures, and official proceedings." Ordinarily, Westergaard went on, the university was not disturbed by newspaper stories, which were "expected to be erratic." But in the present instance he was grieved that those who had "been so generous to the University" should "have been treated ungenerously by a slip in a process that ordinarily functions smoothly."

Westergaard promised Watson that all future publicity would "properly emphasize the part played by IBM and reflect the spirit of the University which is symbolized by its motto Veritas."[1] He then repeated his regrets that the news release had so inadequately presented the

---

1. The text of the news release was revised as shown in appendix A below.

contributions made by IBM. The work of Lake, Hamilton, and Durfee had been "mentioned," he remarked, but still "the great investment by IBM in talent, inventive work, and resources should have been explained much more fully."

So far, so good. But then Westergaard unwisely attempted to sum up the relative contributions of Harvard and IBM:

> Credit could have been given to IBM's employees without detracting from the great role of Howard H. Aiken as originator and as inventor—in one accepted sense of the word—, who conceived the basic theory, chart, and mode of operation of the great calculator and had described these and other principal features of the machine before IBM was approached on the subject.

Though it may seem that Westergaard simply wasn't careful in his choice of words, the presidential archives at Harvard indicate that this was not at all the case. When Conant took Westergaard to task for using the word "inventor" as he had, Westergaard replied that he had done it on purpose.

Westergaard had considered two ways to make amends to Watson and IBM for the contretemps at the dedication. One was to hold what he conceived as a "colloquium" on the calculator, with Lake as the principal speaker and presumably with Watson as the guest of honor. Conant gave tentative approval for this, as did Shapley (who thought it should be called a "conference"). Westergaard believed that Shapley should be the chairman and that the meeting should by jointly sponsored by Harvard's engineering and astronomy departments. Westergaard's other suggestion was to publish a brochure containing texts of all the speeches given at the dedication. Such a publication would stress the importance of IBM and its staff in several separate roles: as engineers, as constructors, and as donor. Watson's name would appear prominently. Nothing came of either suggestion.

On 28 August 1944, more than two weeks later, Watson acknowledged receipt of Westergaard's letter. His reply consisted of three brusque sentences, each a separate paragraph. In the first Watson declared abruptly and formally that he had received the letter, "the contents of which have been carefully noted." In his second he adopted a chilly tone of rebuke: "Your letter clearly indicates to me that you are not familiar with the invention and development of the IBM Automatic Sequence Controlled Calculator." In the third, he wrote that, owing to of the "importance to Harvard University and the International Business Machines Corporation," he "suggest[ed]" that Westergaard "investigate this situation and obtain all of the facts." Then, "Yours very truly."

Between Saturday, 12 August 1944, when Westergaard wrote his unwise letter to Watson, and Monday, 28 August, when Watson replied, Harlow Shapley, who actually had known Watson before the ASCC/Mark I affair, had tried to smooth the troubled waters. On Friday, 11 August, Shapley (who was enjoying a few days of vacation in Peterborough, New Hampshire) received a memorandum from his secretary in Cambridge (one L. Beresnack) informing him that Dean George Henry Chase had telephoned that morning to tell Shapley "that a very important luncheon meeting with the IBM representative" was to be held at the Harvard Faculty Club on Monday, 14 August. Chase had "stressed the importance" of Shapley's participation. Presumably this meeting took place as scheduled, perhaps even before Watson had read Westergaard's letter. Apparently, on the following Friday, 18 August, Walter Lemmon was the target of Watson's anger at Harvard, at a meeting in connection with the tensions between IBM and Harvard. By this time Watson had certainly read Westergaard's infuriating letter. The only source I have been able to find concerning Lemmon's encounter with Watson is a letter from Shapley to Lemmon, dated Saturday, 19 August 1944, that contains this passage:

In spite of my usual calm self-control, I cannot help but be both worried and exceedingly sorry that you ran into temperamental difficulties on Friday. It would be too bad if you personally would have to be penalized as well as humiliated because of your indirect association with this occasionally stumbling institution.

On the Friday of Lemmon's encounter with Watson, Shapley—who was in New York on business related to his collaboration with Lemmon—met at the Century Club with two officials of IBM: Edward (Ned) Douglas, a vice-president, and John Hayward, a lawyer and the editor of IBM's house magazine, *Think*. Of this meeting Shapley wrote to Lemmon the next day: "What an evening! At least temporarily, Conant, and perhaps Shapley, are riding high with T.J.W.; but Howard Aiken is sunk." Shapley felt, as did others at Harvard, that Aiken had "been somewhat persecuted," but Shapley was willing to characterize Aiken as having "definitely been exceedingly difficult, ill-mannered, with his head turned by brass-hat [i.e., Navy] connections of the past three years." Shapley was not yet sure how the situation would be resolved: "Maybe we shall pull out—maybe not." But he was able to work very well with Hayward, and he expected that on 22 August he would be conversing by telephone with Hayward and with others "in the interest of the eternal amity."

But on 28 August, Shapley wrote to Lemmon that the "IBM troubles" continued. This was the date of Watson's curt reply to the communication sent by Westergaard on 12 August. As Shapley later explained, in a letter to Theodore Brown, Westergaard's "unfortunate letter" to Watson had "caused great offense."

In his letter of 28 August to Lemmon, Shapley said that he was going to see IBM's John Hayward, who was then in New York, on the following morning. The next day appears to have been when, as Shapley later wrote, he "spent all of one day on the job in New York." This was the day of his "interview at the end of August with Mr. Watson and his assembled engineers and lawyers," as Shapley later described it to Theodore Brown. Shapley wrote a followup letter to Watson, thanking him for "the opportunity" to be able "to discuss personally with you and your colleagues the complex problems connected with the gift of the great Calculator."

By the time Watson's curt reply arrived in Cambridge, Harald Westergaard was no longer around. Immediately after writing his letter of 12 August to Watson, he left Cambridge "for an extended turn in the Navy." On his departure, he delegated the responsibility for all problems relating to Aiken and the calculator to E. L. Chaffee. Chaffee, director of the Cruft Laboratory, thus became, in Dean Westergaard's absence, "Supervisory Officer of the Computation Laboratory." But Chaffee wisely passed a copy of Westergaard's letter to Watson on to Harlow Shapley, thus leaving all the problems that had arisen in the relations between Harvard and IBM in Shapley's hands. At Conant's request, Shapley had already agreed to this role.

Shapley was an excellent choice. He was one of America's most distinguished scientists, and he had long been interested in machine calculation. In addition, he had been a key figure in bringing Aiken into contact with IBM. He had sent Aiken to Theodore Brown, as he was pleased to remind Watson in a carefully crafted and conciliatory letter.

Shapley's letter to Watson is dated Thursday, 31 August, the day after his visit with Watson at IBM. Obviously this letter was a formal statement for the record. It conveyed, first of all, Dean Westergaard's apologies. Shapley told Watson that Westergaard, even before he received Watson's reply to his letter, and just before he left Cambridge, had gone to Shapley's office "at the urgent request of President Conant" and had learned from Shapley himself "of the incorrect press releases and other phases of the publicity." Westergaard had requested that Shapley "make appropriate apologies on his behalf, since he recognized at once that what you have now written is right—that is, he had hastily written his letter to you without sufficiently knowing the facts." Since he had previously "been incompletely informed and misinformed," Westergaard wanted Watson to know of his regret for "his wrong conclusions."

Shapley told Watson that he spoke also for himself, noting that he wanted "to move as speedily as possible in rectifying the errors that were made." He was also writing to convey to Watson the personal regrets of President Conant, who was "deeply disturbed by the mistakes that were made." Conant, according to Shapley, recognized that the "press statements were misleading and must be corrected" and admitted that there had been a "decided discourtesy to [Watson] and [his] colleagues," constituting an act "unbecoming to Harvard University."

In particular, Shapley wrote, William Claflin, Harvard's treasurer and a member of the Harvard Corporation (Harvard's primary governing body), had "collaborated with" Shapley "in making an official correction of the erroneous statement in the *Christian Science Monitor.*" Moreover, Aiken was "writing letters to Mr. Lake and to Mr. Bryce" and was preparing (at Shapley's explicit suggestion) a "statement that you may want to use in [your house journal] *Business Machines,* not only straightening out the matter of correct credits, but giving a brief report on the current successful operation of the Calculator."

Then, in a master stroke of diplomacy, Shapley assured Watson "most sincerely that Commander Aiken regrets any discourtesies toward the officials and engineers of IBM." According to Shapley, Aiken also "realizes that his nervous, worried, and much overworked condition of the past few months is not sufficient excuse" and that "the publicity concerning the machine was both unfair and wrong."

Shapley concluded on just the right note for Watson. The "new Calculator and the developments that will immediately follow," he wrote, are "epoch-making in the history of applied mathematics." As one who truly was "deeply and objectively interested in the advance of American science," Shapley wanted "personally" to express to Watson his "appreciation of the work of all the men of IBM who have provided this powerful tool for powerful thinkers."

As Shapley later told Theodore Brown, Watson found this letter "very satisfactory." Nevertheless, much work still needed to be done toward healing all the wounds.

On 31 August, Shapley's secretary typed a draft of a letter, presumably dictated by Shapley, intended for Aiken to revise and send to Watson: "As Dr. Shapley has fully explained to you, I regret very much that on the occasion of your gift of the Automatic Sequence Controlled Calculator to Harvard University, many of the news releases were erroneous, and that I was wrongly credited with inventions that had been made by others. I trust that you realize that I have only the best of wishes for IBM and all the employees of the Company with whom I have come in contact." After a statement that Aiken would "write directly to Mr. Bryce and to Mr. Lake," the draft concludes: "With my personal appreciation for your consideration and support during the past six years, I am, Very sincerely yours. . . ."

But Aiken wrote his own letter. It is dated several weeks later: 20 September. I have not been able to locate the actual letter; however, if Aiken's letters to Bryce and Lake are any guide, it is doubtful that he went as far as Shapley's draft had suggested. Aiken was sorry for the unpleasantness, but evidently he *did* consider himself the chief inventor.

On 2 September, a few days after composing the above-mentioned draft, Shapley wrote to Walter Lemmon: "With my tongue in my cheek, and my heart in my boot, and my pride in the cellar, I am actually making real progress in clarifying the IBM-Aiken mess." But on September 5 Shapley had to tell Lemmon: "I continue to sweat blood over the IBM mess; Aiken has blown up, and the Monitor reporter had to be assuaged." Yet he could add this: "Hayward is nice to work with. And in time we shall fix things well enough."

In his letter to Watson dated 31 August, Shapley had not only alluded to the letter sent in Claflin's name to the *Monitor,* he had also enclosed a copy of that "official letter of correction." Shapley later explained to Theodore Brown:

In that letter to Mr. Watson I also sent along a copy of a correcting letter to the Editor of the *Christian Science Monitor,* prepared by Mr. Hayward and me and signed by Mr. William Claflin, as Treasurer of Harvard University. The letter was very specific, and I think satisfactory.

In another letter, composed on 15 September but not transcribed until 29 September, Shapley described the matter at length to Claflin:

Dear Bill:
Thanks for signing the letter to the Monitor. It was composed, as I told you on the phone, by the lawyer, Mr. Hayward of the IBM, and it should go a long way toward appeasing the injured feelings in New York and Endicott. We could not write to the editor of the *Monitor*, of course, that the mistakes were made by his overenthusiastic reporters, who exhibited both ignorance and enthusiasm. But I think that you and Mr. Conant should keep definitely in mind that such was the case. Howard Aiken is not at all at fault with regard to that *Monitor* story.

It was something like this. On the day that the interviews with Boston reporters were held officially by the Harvard Publicity Bureau, the Monitor reporter asked the question something like "You're making a pile of money out of this machine, aren't you?" That somewhat insulted and considerably annoyed Aiken, as it would anyone who has worked so hard for this Harvard Calculator. He replied that he was not making a cent out of it, that his rights had been assigned to IBM.

The reporter drew the wrong conclusions and wrote the offensive story.

Shapley summed up the whole affair this way:

The New York irritation about these various incomplete or misleading press stories is one of the most remarkable phenomena of my experience with homo sapiens. A molehill has been magnified to a galaxy. It has all been pretty hard on Aiken, who has had to bear the brunt of the embarrassment. The clumsiness of two or three others is put on his discredit ledger by IBM, notwithstanding my explanations.

Shapley concluded with an apology for Aiken. We "on the inside," he wrote, "should realize clearly that the *Monitor* distortion and the other erroneous press releases and misstatements were not of his making."

On Monday, 18 September, on board the New Haven Railroad's train "New England States," Theodore Brown penned a request to Harlow Shapley: "Just before leaving for this train, Mr Watson of the IBM called me on the phone [to set] a date to go over my part in the Calculating Machine episode." The date was "set for Tuesday Sept 25th in Endicott." Brown's note to Shapley is dated Monday 18 September. Brown must have erred when he wrote that Watson had set a date of "Tuesday 25 September" for the meeting. Furthermore, a letter from Brown to Watson dated 23 September (a Saturday) refers to his having seen Watson at Endicott the day before. Very likely, the date on the letter—handwritten on a train—is incorrect. Perhaps the letter was written earlier than Monday, 18 September.

Although Brown wrote "it will be wise for me to keep the appointment," he first wanted "to clear with you [i.e., Shapley] or someone the events of the past weeks" so that he could "follow policies already set up." His trip, he wrote, "will keep me most of the week at Fort Leavenworth." He was planning to "return either Friday or Saturday," and he would "phone as soon as I get back." He hoped he would be able to "see or talk with" Shapley on "either Saturday or Monday."

Shapley replied in a long, detailed letter dated 20 September. He began by notifying Brown that he would not be able to see him on the proposed Monday. "It would be impossible in one finite letter," he wrote, "to tell you of the goings-on and the runnings to and fro with regard to the 'episode.'" But Shapley could give Brown "a few highlights, which may be sufficient." Shapley was "dictating this very rapidly before I catch a train" and so he feared it "may not be too coherent." But "possibly you will be able to get an indication of how things are." Shapley said he had "found it very comfortable to work with Mr. Hayward." He had "realized that various factors, such as hot weather, disturbed health, worry, and confusion, have been explanatory contributing factors to the peace of minds of Mr. Watson and his colleagues." In one of the most interesting parts of the letter, Shapley wrote: "There is some difficulty in straightening this thing up, because of the great sensitiveness, and the essentially psychopathic situation in New York and Endicott; and also because Commander Aiken has rather rightly felt that he has been persecuted and abused, etc." Shapley noted that Aiken "regrets the miscarriage of publicity, but . . . declines to go on his knees." Shapley then explained that "Commander Aiken has had to be out of town a great deal, and has been working almost literally nights and days on the Navy job." Shapley added: "You should know, of course, that in his letters and in the article, Commander Aiken has not crawled on his belly and whined for forgiveness." Shapley had "not forced him beyond certain limits in the appeasement policy." Aiken "has not been at fault for some of these things that have gone wrong," Shapley continued, "and the IBM is certainly making a deliberate attempt to belittle Aiken's contribution to this work." Shapley suggested that IBM "may need to do that for their pride, or it may be because of patent interests, or it may be for some other reason I do not know." Perhaps the most interesting part of this letter is this: "Commander Aiken has . . . drafted a letter to Mr. Watson, which I shall try to discuss with Mr. Hayward tomorrow before it is actually sent along." Evidently, Shapley wanted to have Hayward read Aiken's letter in order to make certain that it did not contain anything that might offend or irritate Watson. Shapley characterized Aiken's letter of 20 September as "a friendly and grateful communication."

On 20 September 1944, about a month and a half after the dedication, Aiken wrote a letter to T. J. Watson, another to J. W. Bryce, and yet another to Clair Lake. He opened the letter to Bryce by apologizing for the delay in writing. He explained that he had been extremely busy with "major problems attendant upon starting this project," but now that his working day had been cut down to "a mere twelve hours" he had time to write a "long letter long overdue." He told Bryce how "extremely disappointed" he had been upon hearing that Bryce could not be present for the ceremonies on 7 August—especially since Bryce had been "the first to realize the possibilities in the IBM Automatic Sequence Controlled Calculator on the occasion of our first discussion in November 1937." "I believe you know the high regard I have always had for both you and your work since that date," he added. "It was a pleasure for me to express this regard at the dedication." Since Bryce had not been present, Aiken enclosed a copy of his remarks. Aiken then expressed pleasure that the "calculator exceeds our expectations." Aiken and his group has been able to complete "ten reports of importance," all of which would have led to publications "in other times" but which he could not describe to Bryce because they had to do with the war effort. Aiken looked forward to the end of the war, when "we may resume our peace-time activities and begin making those contributions to Applied Mathematics which are well nigh impossible" without the calculator—"or at least accomplished . . . only at great expense of time and effort." Aiken made no mention of the news release.

In the letter to Lake, Aiken said that he had been looking at some "old notebooks" in which he had recorded the substance of the early conversations about the new machine. He had been reminded of some of "the lighter moments" of his association with Lake. He recalled to Lake their first meeting, when "you and Ham were just a little worried about the responsibility of having a Harvard bloke on your hands, and ultimately announced that you would not hold the association against me permanently." "It is pleasant," Aiken wrote, "to think over these old times" and "unpleasant to recall that the very time when we should have been the most elated with our work was one of mutual misunderstanding and unpleasantness." As far as he was concerned, Aiken added, the "last few weeks comprised but a small part of our acquaintance." He preferred "to recall the years of pleasant association rather than the immediate past." Then, with the optimism that characterized his hopes for the future, Aiken changed the subject. The work that "we started" must "go on," he declared, because "the future of the physical sciences is dependent upon faster and faster, and more accurate computation." This letter is written in a friendly, almost comradely style. Aiken refers in passing to "mutual misunderstanding and unpleasantness," but he does not even hint that there was a serious misstatement or fault in the news release. Nor does he apologize for having been assigned the major credit for the invention. He certainly does not express regret for the distribution of credits in the news release, nor does he even hint that IBM and Lake, Hamilton, and Durfee were not given sufficient credit for their part in designing and constructing the machine. The closest Aiken comes to such a statement is in the penultimate paragraph, where he tells Lake to take care of his health and to "pay us a visit in Cambridge and watch the best Calculator in the world turn out the most accurate results in the least time so far accomplished." "This," he adds, "is the reward of your efforts, and Ham's and Ben Durfee's and mine." When he next sees Lake, Aiken concludes, he will "in person extend his heartiest congratulations."

In a reply dated 3 October 1944, Watson thanked Aiken for his letter and for "your kind remarks about me at the dedication of the Harvard machine." Watson then referred to the letter Aiken had written to Lake, of which he had "just seen a copy." Watson restated his strong feelings about the "original press statements given out, identifying you as sole inventor of the machine" and not giving Lake, Hamilton, and Durfee credit "for their very important and untiring efforts." Watson therefore felt a need to tell Aiken that "it would have been a gracious gesture on your part" and "very much appreciated" by Lake, Hamilton, and Durfee if Aiken's "letter to Mr. Lake had contained an acknowledgment of the sincere regret over such unfortunate and erroneous publicity."

The last three sentences of Watson's letter to Aiken may well be the most interesting. They indicate that Aiken's letter had not dealt exclusively with the news release and the reporting but had also introduced topics related to the machine, such as its speed in performing various numerical calculations. Aiken had sent Watson a memorandum on the computation of logarithms, and this must have raised the question of patents. "Undoubtedly the scheme is not patentable," said Watson;

however, if IBM could "work out a structure to use this scheme of computation," and "if it turns out to be novel," then that "may be patentable." He closed, affably, with "best regards."

Both Watson and Harvard were anxious to correct the fundamental errors in the Harvard news release and in the newspaper stories based upon it. Harvard revised the news release.[2] Watson, not content with that, enlisted his own publicity department to change the public image of Aiken as sole inventor of the ASCC and to make certain that IBM's contributions to the great machine were spelled out precisely and made generally known. According to the history by Bashe et al., Watson "personally oversaw the preparation of a carefully worded brochure," which began with the statement that the Automatic Sequence Controlled Calculator was "the latest advance in International Business Machines Corporation's program of adapting its equipment to use in the field of scientific computation."

Denying Aiken's primary claims as originator, the IBM brochure noted that "Harvard University's need for a machine such as the IBM Automatic Sequence Controlled Calculator had long been a topic of discussion among members of the Harvard Faculty." In the "course of one of those discussions, originated by Dr. H. H. Aiken," the "fact that IBM standard equipment for some time had been successfully used for scientific calculation purposes was brought out by Professor Harlow Shapley, Director of the Harvard College Observatory, and by Professor T. H. Brown, Professor of Business Statistics at Harvard and a consulting member of the Faculty of the IBM Department of Education since 1928." Neither of these sentences is strictly accurate.[3]

The IBM brochure quoted extensively from Aiken's speech at the dedication, including the following paragraph:

It is now eight years since we began thinking in terms of automatic calculating machinery, especially designed for our purposes, and after a preliminary period in which the mathematical theory alone of such machines was under consideration—a period in which the support of Professor Chaffee, Dean Westergaard, and Professor Shapley aided this project—we approached the

---

2. See the footnotes to appendix A.
3. Watson's continuing wrath may have been responsible for the statement in the IBM brochure that the ASCC was "invented by engineers of IBM" and that therefore "full credit for the invention should go to four engineers of IBM: Clair D. Lake, Frank E. Hamilton, Benjamin M. Durfee, and James W. Bryce." Aiken was credited only with providing "the basic theory."

142   Chapter 15

International Business Machines Corporation and asked their support to build such a machine and construct it and put it into operation.

The next quoted paragraph mentioned that Aiken's "first contact with that company was with Mr. J. W. Bryce," who "for more than thirty years has been an inventor of calculating machine parts." Aiken acknowledged that Bryce had "fostered and encouraged this project" and that "the multiplying and dividing unit included in the machine is designed by him."

As quoted in the brochure, Aiken then described how, on Bryce's recommendation, "the construction and design of the machine were placed in the hands of Mr. C. D. Lake, at Endicott." He added that "Mr. Lake called into the job Mr. Frank E. Hamilton and Mr. Benjamin M. Durfee, two of his associates." Aiken described the early days of the design job as having "consisted largely of conversations—conversations in which I set forth requirements of the machine for scientific purposes, and in which the other gentlemen set forth the properties of the various machines which they had developed, which they had invented, and based on those conversations the work proceeded until the final form of the machine came into being." IBM's official commentary on this allocation of credits reads as follows:

> The early conversations between Dr. Aiken and IBM's engineers having disclosed that the known requirements could be met by a combination of some of the company's standard and special mechanisms, such as those used at Columbia and other universities, IBM proceeded to build the machine.

A list of "the more important" of "the many basic units of the Calculator, invented or developed by IBM engineers" follows.

The above-mentioned list in the IBM brochure was repeated, word for word, in a letter written by Paul H. Buck, then Provost of Harvard University, that appeared in the *Wall Street Journal* of 14 January 1948, That letter was a response to an article in the *Journal* that had given "less than due credit" to IBM for the ASCC. In this letter, the list taken from the brochure was preceded by the following paragraph:

> The original IBM automatic sequence controlled calculator, which was formally presented to Harvard University by IBM in August 1944, was invented by engineers of IBM following the basic theory of Professor Howard H. Aiken of Harvard. Full credit for the invention should go to four engineers of IBM: Clair D. Lake, Frank E. Hamilton, Benjamin M. Durfee and James W. Bryce. The machine was built to their specifications after six years of laboratory work in the IBM plant at Endicott, N.Y., with Mr. Lake in charge of the inventive work.

Note that the provost has now reduced Aiken's contribution to "basic theory," and that "full credit for the invention" is said to be belong to the "four engineers of IBM."

Upon close reading, Aiken's speech at the dedication does not deny the contributions made by IBM, but it does stress Aiken's initiative. The unsolved question is why the Harvard news release, and consequently the press reports, downplayed IBM's part in the invention so greatly. No doubt Aiken was inexperienced in dealing with the press (even the Harvard News Office), and certainly he was carried away by the very success of the machine.

Aside from seeming to have claimed the lion's share in the new invention and to have assigned the IBM engineers a role inferior to his own, Aiken, Watson would have felt, had slighted him personally. Watson's authorized biographers, T. G. Belden and M. R. Belden, make this absolutely clear, even to quoting Watson as telling Aiken that "most of the ideas in IBM have come from me personally." Belden and Belden add that Watson's "name appeared on many [patents]" on "the same basis now claimed by Aiken"—that is, on "the basis of an original suggestion."

During our interview with Aiken in 1973, Hank Tropp and I carefully refrained from introducing the topic of Aiken's relations with Watson or with IBM, and his views of the relative contributions to the machine made by him and the IBM engineers. This would surely have led to a heated conversation, which would have deflected us away from more basic considerations of computer history. We were saving the story of Aiken's relations with IBM for a second session, which never took place. But it was clear from many disparaging remarks Aiken made about Hamilton and Durfee and about IBM in general that the hostility generated at the time of the dedication was still strong. Only for Bryce did Aiken show any affection or respect.

Frederick P. Brooks Jr., one of Aiken's most distinguished students, recalls that Aiken discussed his relationship with IBM years later during a coffee hour in the Comp Lab. "It was a source of deep bitterness on his part," according to Brooks, "and between him and Tom Watson Senior." Brooks had "had an opportunity to discuss it with Tom Watson Junior," and "it was clear that the feeling was strong on both sides." In Brooks's opinion, "the particular incident that triggered the greatest distress" was the "handling of the press" at the dedication. Brooks believes that Watson Sr. saw the news release and the interviews Aiken gave to the press as elements of a "deliberate" campaign on Aiken's part to get most of the credit for the new machine.

Brooks also suggests that there is "a technical reason for some of the misunderstanding." At the time, the distinctions among "the architecture, the implementation, and the realization of computers" were not yet understood.[4] In retrospect, Brooks finds that "the architecture and what we today would call a good bit of the implementation" of the ASCC/Mark I were "designed by Aiken." The "rest of the implementation" and "all of the realization" were "designed by Durfee, Lake and Hamilton." In short, according to Brooks, if the "distinction between those roles" had been understood at the time of the dedication, "some of the bitter emotional feeling over the question of credit" would probably have been "defused."

A similar point of view appears in *IBM's Early Computers*. Referring to Paul Buck's letter to the *Wall Street Journal,* the authors emphasize the occurrence of the phrase "basic theory" in the statement that the ASCC/Mark I "was invented by engineers of IBM following the basic theory of Professor Howard H. Aiken." That is, the phrase "following the basic theory" may seem to "attenuate Aiken's contribution," but it was "instructive for the time." Some 15 years later, Bashe et al. note, "IBM employees would invoke the term 'computer architecture' to embrace register formats, storage capacities, machine instructions, and other matters of direct concern to the system user. Then, in retrospect, Aiken could have been described reasonably well as the ASCC architect."

Computer architecture is a concept that has been changing since its inception. In 1962, in his celebrated article on "Architectural Philosophy," Fred Brooks described it as the "art of determining the needs of the user of a structure and then designing to meet those needs as effectively as possible within economic and technological constraints." This definition neatly defines Aiken's fundamental contribution to the ASCC; note that it does not include any specification of technological elements.

I don't know who first introduced the concept of computer "architecture," but as early as 1959 "system architecture" was used by Lyle R. Johnson at IBM in "A Description of Stretch." At this time, "system architecture" was used "to suggest a level of computer structure that subordinates details of logical and circuit design." This concept specifies Aiken's contribution to the ASCC/Mark I. This function was referred to in Aiken's teaching philosophy as "design." In the

---

4. Brooks says that he and Gerrit Blaauw (a fellow student at the Harvard Comp Lab) originated such distinctions.

Students' Handbook for the Automatic Sequence Controlled Calculator, Mark I (1949), Aiken's point of view is presented as follows: "the 'design' of a . . . computing machine" is "understood to consist in the outlining of its general specifications and the carrying through of a rational determination of its functions." Aiken's concept of "design" is so well expressed in this document that it is worth quoting a second time. Design, according to this philosophical position statement, "does not include the actual engineering design of component units." When I encountered this statement, it seemed to me that Aiken was actually specifying his role in the ASCC/Mark I—producing the "design."

# 16
## Some Features of Mark I

Nowadays, the word "computer" invokes the image of a desktop machine about the same size as the cash registers that Thomas J. Watson sold as a young man. However, each of the early computers or giant calculators,[1] such as Mark I or ENIAC, occupied a large room.

Most accounts of Mark I stress its huge size and the enormous number of parts (765,299, according to IBM's official records, including 3300 relays and 2200 counter wheels). Weighing 9445 pounds and containing 530 miles of wire, it was the largest and most complex electromagnetic device that had ever been constructed. About 50 feet long, 8 feet high, and almost 3 feet wide, and sheathed in stainless steel, it was an imposing sight.

As figure 16.1 shows, Mark I consisted of a series of panels in a straight line and two supplementary panels jutting out at 90 degrees. Panels containing a subsidiary mechanism were added later. The straight-line configuration allowed a main drive shaft to run the whole length of the machine.

Panel 1 contained 60 constant registers, each consisting of a row of 24 ten-position dials. These held the constants that appear in any algebraic or differential equation. A typical problem would also consist of terms such as variables (which could be raised to powers) and differentials of several orders, each of which might have a constant coefficient. At the beginning of each problem, an operator would enter the constants into these two panels. By means of dial switches, each of which could be set by hand to a position corresponding to 0, 1, 2, 3, 4, 5, 6, 7, 8, or 9, a number with up to 23 digits could be entered into each register; as has already been noted, the 24th place was reserved

---

1. See appendix H below for a discussion of whether Mark I, ENIAC, or any of the other early machines fits the present-day definition of a computer.

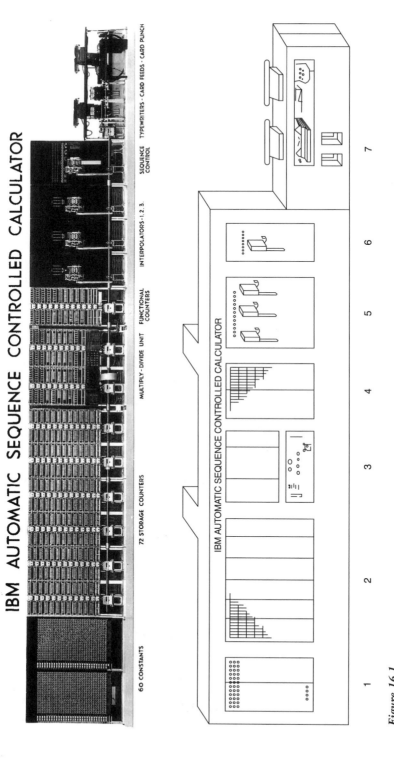

*Figure 16.1*
The functional elements of the ASCC/Mark I.

to indicate whether the number was positive or negative. Since the location of each of these 60 registers was assigned an address, the instructions could use the location to identify a data number being called up in the course of a calculation. The programmers had a "code book" listing the address that designated each register and also assigned operation code numbers to the various steps the machine would follow in solving a problem.

Panel 2 contained 72 registers which were "accumulators." Each of these was made up of 24 electromagnetic counter wheels—again providing the capacity for 23-digit numbers, with one place reserved for sign. This second set of panels constituted the "store" or "storage." A typical line of coding would instruct the machine to take the number in a given register (either a constant or a number in the store) and enter it in some designated register in the store. If there already was a number in that register, the new number would be added to it.

A separate unit was provided (panel 3) for the multiply/divide unit. Panel 4 contained the functional units. They had the capacity to convert a number into a base-10 logarithm, an antilogarithm, or a sine. Panel 5 held three tape readers, which provided the machine with a capacity for interpolation. Panel 6 contained the sequence-tape reader, described as the heart of the machine since the tapes fed into it contained the program to be executed, step by step.

At the end of the machine were two punched-card feeds. There was also a card punch and a pair of electric typewriters, which typed out the results of computations (usually in tabular form).

Numerical inputs could made by setting dial switches or by reading punched tapes or cards. Output could be in the form of typewritten tables or decks of punched cards. The shaft, running along the base, was always turning. Relay circuits were introduced in order to activate the clutches that would link the wheels of the registers with the turning shaft. This action was synchronized so that a counter wheel would advance only through the number of positions designated by the stage of the calculation. For example, the clutch might connect a given register to the shaft for seven units of time, in which case the number 7 would be added to whatever number had been in that register at the start. This kind of addition required, on average, about a third of a second.

Mark I was especially adapted to tabulation of functions. That is, it could produce tables of the values of functions for different values of the "argument" or independent variable. Of special importance in such work was the ability to compute values of the function for regularly incremented values of the argument. But Mark I could also be programmed to solve ordinary and partial differential equations and systems of linear equations. It is often mistakenly said that Mark I was essentially a machine for producing numerical tables, but an examination of the problems actually assigned to it during and after World War II shows that Mark I was able to deal with a large variety of problems. This misunderstanding may have arisen from postwar use of the machine to generate a vast collection of tables of Bessel functions.

Mark I embodied a twentieth-century variant of an important innovation in making tables that had been suggested in the nineteenth century by Charles Babbage. Babbage, as has been mentioned, was concerned to eliminate the kinds of errors that always tend to be introduced whenever tables are copied by hand and set into type (as in typesetting and proofreading). Accordingly, he designed both his Difference Engine and his Analytical Engine to stamp out results on a papier-maché moulage, from which a stereotype plate could then be cast. Mark I used two IBM electric typewriters to type out its results in tabular form on sheets of paper. (Aiken often complained that the IBM typewriters were much slower than the computation.) The printouts were then placed in order in specially prepared forms, producing copy for each page. These pages were then photographed, and the text was "burned into" special zinc plates for printing.

In retrospect, the most significant design feature of Mark I was Aiken's idea that the sequence of mathematical operations should be controlled automatically by a program entered (or coded) into the control tape. One of the advantages of this was that an operator could be working on programs for future problems and checking the sequence of coded instructions while the machine was running other programs.

While tables were being typed out, an operator would watch the numbers as they appeared, looking for a sudden large jump between one value and the next (which would signal an error). This was one of the many ways in which Aiken arranged for the accuracy of the calculation to be monitored. Aiken always tried to have some feature of each program designed to check the accuracy of the calculation in progress. Thus, if a sine was called for, the values could be checked by computing the cosine and seeing whether the sum of the squares of the two equaled 1.

Anyone who ever witnessed the operation of Mark I will remember the pleasant whir of its main drive shaft. In the 1950s, when Aiken had an office near the machine, he usually kept his door open so as to be able to hear this sound. I well remember being with him one time when he detected a variation in the sound. He leapt out of his seat, ran to a switch, and turned the machine off until the fault could be located and corrected.

One major difference between Mark I and IBM's business or accounting machines was the use of punched tape rather than punched cards. Since the tape was made of IBM card stock, and since the mode of operation was somewhat the same for both, it is sometimes mistakenly supposed that Mark I was essentially card-driven, the cards being joined together. This is not quite the case, however, since the tape provided continuous reading of numerical data or instructions, whereas cards joined together would still contain separate groups of such numbers and instructions on each card. As Aiken explained during the 1973 interview, cards are easier to edit or sort, but they can also invite disaster—especially when being transferred. Old computer hands are likely to have memories of trays full of punched cards spilling all over the floor. Individual cards could be bent or damaged. Many readers will remember the warning on punched cards not to "fold, spindle, or mutilate" them. Aiken's high standards of accuracy and reliability were better served by continuous rolls of punched tape than by trays of punched cards.

As has been mentioned in an earlier chapter, Mark I was constructed of the types of components—relays, counters, card readers, card punches, and switches—that had been used for many years in IBM's business and statistical machines. Nevertheless, many of these components were new in design and had features that were important to the successful functioning of Mark I. For example, the new relays and counters were much smaller and faster than previous ones. It is difficult to imagine how Mark I would have functioned at all if it had had to accommodate the older relays; it surely would have been even slower than it was.

Mark I's built-in sequences of operations or functions (including base-10 logarithms, sines, exponentials, and interpolations) were officially designated in the Manual of Operation as "built-in subroutines." These were very much like subroutines in the sense in which we know that term today.

The early programmers of Mark I—primarily Richard Bloch, Robert Campbell, and Grace Hopper—would reduce each problem to a program consisting of a sequence of mathematical steps or mathematical operations. They would then turn to the code book in order to find the code designation for each step of the program, producing a line-by-line translation of the mathematical steps of the program into numerically coded statements that the machine could read, switch settings for the constants registers, and so on. The mathematician or programmer worked with the code book at his or her side. Before long, it was evident that certain sequences of the instructions occurred again and again. So the practice began of writing such partial programs into a notebook. The most extensive such collection was assembled by Bloch, the primary programmer.[2]

Grace Hopper informed me that she and others kept private libraries of partial programs. Years later such collections of partial programs became known as libraries of subroutines. Their first use can be traced back to the programmers who worked on Mark I. (These coded subroutines, or "canned" elements of programs, are to be distinguished from the subroutines built into the machine, such as those for logarithms, exponentials, and trigonometric functions.)

As originally conceived, Mark I had no conditional or branching circuits. These were added later. As was noted in chapter 8 above, the absence of branching circuits in the original Mark I was a sign of Aiken's very restricted knowledge of Babbage's writings. In fact, even when branching circuits were added to Mark I they were of a very limited nature.[3]

Other later enhancements of Mark I included a subsidiary sequence for subroutines, an electronic multiply-divide unit, additional storage registers, and a second sequence tape unit. Although Mark I operated with a fixed decimal point throughout its long life, the position of the decimal point was adjustable between problems.

As was to be expected with so complex a machine, there were many problems in the early days. Bob Campbell particularly remembers the inevitable errors in writing programs and also in setting up prob-

---

2. Aiken told Tropp and me that Bloch was so skilled that he would write programs in ink.

3. For details see appendix D.

lems—notably errors in the inputting of constants. At first there were also operator errors, often due to imperfect instructions. In the early months, however, the greatest problem was that the machine was not very reliable and continually made errors. The worst feature of these errors was that they were "intermittent" and seemingly random. As a result, they were very hard to trace down to a source—and while the source of one error was being tracked down, another error might suddenly pop up. Eventually, most of these errors were traced to the relays and the wiring contacts.

Clair Lake of the IBM engineering staff designed the relays used in Mark I so that they could be readily unplugged and replaced. Many of the wires connecting the mounts were crimped (or "swedged") rather than soldered. Because many of these wires were crimped only once, many of the electrical contacts were poor. This troubling source of unreliability was eliminated over the course of several months during which all the singly crimped connections were replaced by doubly crimped ones.

The relays presented a different and far more serious problem: they produced errors that could not be reproduced and so traced to their ultimate source. It is usually believed that Aiken "chose" relays as Mark I's primary components on the basis of their reliability; yet these same relays proved to be a primary source of unreliability. The problem arose because the relays used were not the tried and true ones that had been used for years on IBM electromechanical business machines; they were a new type, described by Bob Campbell as an "inadequate design."[4]

The original relays in the machine made use of "piano wire" transfer contacts, which opened and closed a circuit by making contact with brass terminals. This kind of contact produced a very high resistance, and the resistance increased with continued use, often reaching a point where the relay ceased to function properly. Even after this was discovered, Campbell reports, it proved difficult to "adjust the relays to provide enough 'wipe' of the contacts to prevent contact resistance from building up." In some cases, this could be remedied by cleaning and readjustment. Many relays, however, had to be replaced. It was

4. In an earlier publication I suggested wrongly that the relays failed because they had not been designed or built to operate 24 hours a day. In a critique of that statement, Herbert Grosch correctly pointed out that the continuous operation of the machine did not exceeded the normal load imposed on the relays in the operation of IBM's accounting and tabulating machines.

only after a year of these improvements that the machine finally reached its celebrated high level of reliability. Eventually, the machine was reported to be "routinely producing useful results" over 90–95 percent of full-time operation.

The relays designed by Lake were not the only new components used in the ASCC/Mark I. It is often mistakenly assumed that the IBM engineers constructed the machine out of standard off-the-shelf parts and standard circuits. This was not at all the case, although at first, when the plan was to have the machine built at Harvard, it was envisaged that IBM would furnish "standard units such as counters and relays."

The Lake relays were extremely important for several reasons. First, they were ten times faster. Second, they were considerably smaller than those used in IBM's statistical and business machines. If traditional off-the-shelf relays had been used, Mark I would have been even more gigantic than it actually was. Third, they were pluggable, so any single relay could easily be replaced if there was a malfunction.

In November 1940, Aiken reported to Lake that he had shown the new relay (along with other components) to members of the Harvard Faculty, who sent their praises to Lake for his invention.

Among the other novel components was a new sort of counter. There were also new circuits, many designed by Hamilton and Durfee. The circuit used to produce division was devised by Bryce.

Hans W. Spengler of the House for the History of IBM Data-processing calls special attention to the following new components that had not at that time been used in IBM business or statistical machines: the wire-contact relays developed by Clair Lake, the small counters (or registers), table-tape readers that could read in either forward or reverse directions (which the standard read brushes could not do), contact cams operating at twice the standard speed (300 ms/cycle). He concludes that these were not "standard parts which could be taken from the shelves." Rather, "they had to be developed, tested, and produced" for the ASCC/Mark I. Of course, as Spengler notes, Lake used some of these new components soon afterward in his design of IBM's PSRC for Aberdeen. And after World War II these became "parts of IBM's standard elements."

It is thus entirely incorrect to say that the components of the ASCC/Mark I were identical to those then in current use, or that the machine was an assembly of off-the-shelf standard IBM parts and circuits—simply a kind of "super" IBM business machine.⁵ It is true, however, that the construction elements of the calculator were of a traditional type. That is, the primary operating elements were electromechanical, not electronic—relays and counters, as in standard IBM machines, not vacuum tubes.

That many of Mark I's components and circuits were not the standard ones used in IBM business machines has many implications of real significance for an understanding of the history of that machine. For example, as the historians of IBM have pointed out, the machine would be of some practical use to IBM, even though everyone at IBM was aware that it was not the inaugural step in a new product line. As has already been mentioned in this volume, a significant use envisaged for the ASCC/Mark I was to test new components for future IBM machines. Hamilton specifically recalled that the ASCC/Mark I would provide a practical test for Bryce's new circuit for division. Furthermore, because the ASCC/Mark I used so many new components and circuits, everyone at IBM considered it to have been "invented" by IBM engineers. A prominent feature of IBM's booklet about the machine was a long and detailed list of its novel elements and circuits.

In his lectures at the Moore School in July 1946, Aiken discussed the basic principles of the programming and operation of the Harvard machine and also called attention to some of its shortcomings. The Harvard calculator had but one sequence mechanism, Aiken told the students, but "at least four sequence mechanisms should be provided for a large scale calculator." He explained at some length how the greater number of such mechanisms would save machine time and simplify the coding. He referred to the incorporation of "60 constant registers" when "20 would have been sufficient" as a significant "error made in the construction of the machine." He explained the importance of this as follows:

Since our calculator has an accuracy of 23 digits and the algebraic sign, each constant register contains 24 switches, making a total of 1440 switches all told. The time required to set up so many switches is prohibitive, to say nothing of

---

5. IBM itself has fostered this point of view. In the booklet issued by IBM describing the new machine (see chapter 15 above), it is explicitly said that a decision to go ahead and build Aiken's machine was based on the recognition that Aiken's "requirements could be met by a combination of some of the company's standard and special mechanisms." The latter were "such as those used at Columbia and other universities."

156    *Chapter 16*

the errors introduced thereby. When a problem requires input data or parameters consisting of twenty or more numbers, it is far better to introduce them either by means of the interpolators or the card feeds.

Aiken's proposed alternative would also have the "advantage that preparations for a new problem can, in large part, be completed without stopping the calculator itself." Furthermore, "the tapes and cards may be stored for future use with the same or similar problems," but "any data set up in a switch is irrevocably lost when the setting of the switch is altered."

Aiken told his Moore School audience that during "the past two and one-half years our calculator has operated on a twenty-four hours a day, seven days a week schedule." From this experience, he commented, "we have learned that reliability is the most important characteristic of a large scale calculating machine." Mark I, according to Aiken, "operates successfully about 90 percent of the time, and has on occasion run for as long as three weeks without an error." Yet, "in order to insure the accuracy of our computed results, they are always checked by the machine before final printing." Aiken concluded his lecture by expressing the hope "that future machines will have an even greater reliability in order that the operating staff may devote as large a portion of their time as possible to the preparation of problems and the publication of solutions rather than to the maintenance of the calculator itself."

From the very start, Aiken wanted to have a real sense of the speed of Mark I, not merely the time and number of cycles each operation required.[6] In the Manual of Operation, Aiken provided two gauges of the speed of Mark I. One was a simple table. He noted that all the times given in that table except those for addition and subtraction could be shortened "by reducing the accuracy of the computation." The other way to gauge the machine's speed was to "solve a problem first by manual methods and then by use of the machine." Such a test or "estimate" had been made, and "apparently the machine [was] well

---

6. The maximum operating times of the functions of Mark I were as follows: 0.3 second to add, subtract, transfer, or read a card; 6 seconds to multiply; 15.6 seconds to divide; 89.4 seconds (2948 machine cycles) for $\log_{10} x$; 65.4 seconds (218 machine cycles) for $\exp_{10} x$; 60 seconds (99 machine cycles) for $\sin x$. The times could be reduced by carrying out such operations as add, subtract, or transfer while the machine was going through the lengthy multiply or divide process. Subtraction was performed by complement arithmetic.

nigh one hundred times as fast as a well equipped manual computer" (that is, a person doing computations by hand, aided by printed tables and standard machines). This estimate, by itself, was not really meaningful, Aiken explained, since "a [human] computer can work little more than six hours a day before fatigue causes him to produce a prohibitive number of errors." Hence, Aiken concluded, the new machine, "operating on a twenty-four hour schedule," was able to "produce" in "a single day" as "much as six months [of the] work" done by a human "computer."

Reviewing the Manual of Operation in *Nature*, L. J. Comrie took issue with Aiken's conclusions concerning speed. He believed that Aiken's examples, "perhaps chosen for their simplicity," did "not do the machine justice." The reason Comrie gave is that the examples could be done "almost as quickly, and certainly more economically, with a Brunsviga or a National." The Brunsviga and the National were two of the standard business or accounting machines that Comrie had adapted for scientific computing. I have read this statement again and again, and I am still not certain of its meaning. The simple and obvious reading is that "the machine" refers to Aiken's new machine. Thus Comrie would be saying that Aiken's "examples" do not "do the [new] machine justice," since the examples were "perhaps chosen for their simplicity." This would imply that the new machine was even more powerful than Aiken had suggested, being able to do more computing in a day than a human "computer" could do in 6 months, even with a National or a Brunsviga business machine. But in that case, the concluding portion—where Comrie referred to problems being solved "almost as quickly" and "certainly more economically" with a Brunsviga or a National—would make no sense whatever. On a second reading, Comrie meant that Aiken had not done justice to computing with conventional machines. In this case, Comrie would have intended that "the machine" be understood as applying generically to business and statistical machines then in general use and not only to Aiken's machine. In this case, Comrie was saying that problems could be solved "almost as quickly and certainly more economically" with a Brunsviga or a National than with the ASCC/Mark I, not that Aiken didn't do "justice" to his own "machine." That is, Comrie's ambiguous sentence may have been intended to mean that Aiken's examples, "perhaps chosen for their simplicity," do not "do . . . justice" to the method of computing with a desktop or other form of standard or business "machine" such as the Brunsviga or the National.

Comrie's review displays a curious shortsightedness with regard to the possible uses of the ASCC/Mark I. "The question naturally arises," he wrote, "Does the calculator open up new fields in numerical and mathematical analysis, especially in such pressing problems as the solution of ordinary and partial differential equations, and the solution of large numbers of simultaneous linear equations?" Comrie found it "disappointing to have to record that the only output of the machine of which we are informed consists of tables of Bessel functions." Producing these, he noted, was "not difficult (to the number of figures required in real life) by existing methods and equipment." Comrie concluded that "to justify its existence" the new machine would have to be used to "explore fields in which the numerical labour has so far been prohibitive."

Comrie seems to have been misled by the fact that the Manual of Operation did not refer to a wide range of specific problems in engineering, science, and even the social sciences, such as were alluded to in the original proposal. Mark I—while ideally suited to the production of tables—was also a powerful tool for the solution of problems such as those listed by Comrie. For example, in Wassily Leontief's research on input-output economics, which later gained him a Nobel Prize, the ASCC/Mark I easily solved systems of 10 or 15 simultaneous equations—a task that all but defied ordinary human capacity with standard desk calculators or business machines.

# 17
## Programming and Staffing, Wartime Operation, and the Implosion Computations

One of Aiken's students in 1949–50, Jack Palmer, remembers his experience in programming (coding) Mark I, which he found to have been very "similar to coding . . . stored-program computers." After the programmer wrote down sequences of codes for addresses and operations in the appropriate columns of a coding form, these codes or their equivalents would be punched into a program tape. According to Palmer, this was "a strikingly similar procedure to that which programmers who programmed the early stored-program computers did when they were programming in machine language." They too "needed to know the codes for the addresses of the words in storage and the codes for the operations to be performed," which they would write down "in a strict instruction format on a coding form" and which would later be punched on cards. In both cases, the numerical codes were obtained from a coding book.

Indeed, to a programmer Mark I was in many ways much like a modern computer—perhaps more so than some other early machines. Mark I differed fundamentally from later computers not only in its very low speed relative to electronic machines such as the ENIAC but also in its initial lack of conditional branching. More important, data and instructions were completely separate. The latter feature, more than the choice of electronics over electromagnetic relay systems, was and remained central to Aiken's thinking about computers. He always insisted on maintaining the integrity of data and instructions.

Aiken did, of course, come to appreciate the power and flexibility of stored-program computers. Indeed, in 1956 he had a UNIVAC installed in the Harvard Computation Laboratory. Aiken, however, was never willing to accept one feature of the stored program: the ability to modify instructions during the course of calculations.[1]

Almost all of the original programming of Mark I was done by Robert Campbell. The first staff consisted of three individuals. Campbell served as analyst, programmer, and operator—part time, since he continued to teach in Chaffee's pre-radar course. David Wheatland was the assistant programmer and general assistant. Robert Hawkins, employed full time on the machine project, was in charge of the actual operation and maintenance of the machine. E. L. Chaffee delegated the immediate overall supervision to Harry Mimno.

Mimno quickly raised the question of the best name to give the machine. Campbell recalls that Mimno wanted to call it a "supercalculator" in analogy with the radio term "super-heterodyne." Within a short time, the question became irrelevant when Watson ordered the stainless steel sheath with IBM AUTOMATIC SEQUENCE CONTROLLED CALCULATOR emblazoned on it.

The first two problems assigned to the machine came from Ronald King, an associate professor of applied physics, and James Baker, a junior fellow in astronomy.[2] Their problems were "well formulated," Campbell recalled. Campbell programmed the problems; then Wheatland and Campbell conducted trial runs, debugged the program, and checked the results of the trial runs before running the real problems.

The first problem, known as Problem A, was King's. Essentially, King wanted to evaluate a "sine-integral" function needed in computing "the mutual impedance between two parallel coupled wire antennas." These results were used by King and an associate in an article published in the June 1944 issue of the *Journal for Applied Physics*, which appears to have been the first publication to make use of results

---

1. Though Aiken rejected this elegant way of dealing with program modification, he did introduce a limited kind of branching circuits into his machines.

2. A third early problem ("problem C") was undertaken for the US Navy's Bureau of Ships. Basically, it involved assigning various batches of steel to the best uses. Every batch had some "uncontrolled impurities" that affected the mechanical properties of the metal. Four such properties (including Young's modulus and the tensile strength) were correlated with ten commonly found impurities. The computed correlations could then be applied to any given sample in order to determine, from the number and concentration of the impurities, what the steel's mechanical properties and, accordingly, its most efficient use. This problem involved solving ten simultaneous linear equations and "calculating multiple regression coefficients and multiple correlation coefficients."

obtained on a computer. These first calculations were for a limited number of cases. A more nearly complete table was soon produced; however, by that time the machine had been turned over to the Navy, and (as was mentioned in chapter 12) King had no Navy clearance.

Baker's problem ("problem B") required the tracing of skew rays in a seven-element (fourteen-surface) telephoto lens with a focal length of 40 inches. Baker, who was designing the lens for manufacture by the Perkin-Elmer Company on behalf of the Army Air Corps, needed to know how the imaging characteristics would be affected by variations in certain primary parameters. A little later, soon after Aiken returned to Cambridge to take charge of his computer, Jim Baker received a call from Aiken, who asked whether he had "anything to have computed." According to Baker's recollection, Aiken and company "had done a lot of in-house calculations and wondered whether they could try something from outside." Baker told Aiken that indeed he had some work that seemed especially well suited to the machine: some additional calculations of ray tracing for the design of the telephoto lens he was devising.

One of the illustrations shows Aiken, Baker, and Campbell discussing this problem. Aiken, who arrived in Cambridge from Yorktown in May 1944, appears in uniform. Campbell, who did not join the Navy until July 1944, is seen in civilian clothes. Baker explained to me that the calculations, especially those arising the tracing of skew rays, had proved "difficult and tedious." He and his group had calculated not "more than two or three of these by desk calculators," and it had taken "hours and hours per ray." According to Baker, "Aiken had me come over, and I sat down and wrote out and developed a set of equations which I thought would be suitable for the machine." Baker remembers finding Aiken cordial and cooperative.

In June 1944, with Commander Aiken in charge of the computer under Navy auspices, Campbell gave up his teaching obligations and became a full-time member of Aiken's staff. In July he enlisted in the US Naval Reserve with the rank of ensign. Ensign Richard Bloch, USNR, had joined the staff in April. Soon a civilian secretary, Ruth Knowlton, was hired. In July, Lieutenant Grace Hopper, USNR, who had been a professor of mathematics at Vassar, joined the staff. Lieutenant Commander Hubert Arnold, USNR, also a mathematician, was assigned to the project shortly thereafter. Four enlisted men were added to the unit, all of them "Specialists (I) First Class." The "I" in this designation meant that they were skilled in the use of IBM equipment. Campbell and Bloch acquainted them with the running of the machine, Hawkins with its operational and maintenance problems. They became the day-to-day operators of the machine, while Hawkins continued to have primary responsibility for its maintenance and its smooth operation. Frank Verdonk, Yeoman First Class, took over the problems of Navy administration. Among the others who joined the staff were four Naval Reserve officers: Lieutenant Edmund Berkeley (who later became well known for his popular books on computers and for his innovations in applying computers to the problems of the insurance business), Lieutenant Harry Goheen, Lieutenant (jg) Brooks Lochhart, and Ensign Ruth Brendel. William A. Porter, CEM, was put in charge of the four machine operators.

In a special category was the mathematician Robert Rex Seeber, a civilian employee of the Navy who was transferred to the computer project in late 1944.[3] Aiken took an immediate dislike to Seeber, which Seeber reciprocated by taking an equal dislike to Aiken. According to Campbell, Seeber "rubbed Aiken the wrong way" when he mentioned his desire to take some vacation time owed to him before joining the computer effort. Aiken needed his help and insisted that he go to work right then and there.

Many years later, in the course of an interview by Larry Saphire, Seeber expressed strong feelings about Aiken's not having given enough credit to IBM. But he also noted that Aiken had been "a driver" who "particularly felt the pressure of the war and the necessity for getting on with the job." As a result, Seeber had gone to work straightaway, "many times working 80 or 90 hours a week," spending "most of [his] time programming." Seeber found, he said, that "his aim in life must be to work for IBM and help them build a bigger and better computer than the Mark I," a computer "that IBM could have for its own use."

During his stint at Harvard, Seeber became aware of the planning for the second machine, Mark II. He made some suggestions to Aiken,

---

3. After being graduated from Harvard in 1932, Seeber had gone to work in the actuarial department of the John Hancock Mutual Life Insurance Company, where he had become expert in the use of desk calculators and IBM punched-card equipment. During most of World War II, he worked for the Navy as a civilian, using IBM machines in pioneering operations research. After reading about the unveiling of the ASCC/Mark I, Seeber had asked the Navy to transfer him to Cambridge to work on the new machine. As soon as the war was over, Seeber left Harvard for a job at IBM.

who "fortunately . . . did not accept these new ideas." Why was this fortunate and "a happy result"? When Seeber left Harvard, he "was able to bring these ideas to IBM, to include them in the Selective Sequence Electronic Calculator," which he "participated in building" after he "joined IBM." After helping to design the Selective Sequence Electronic Calculator, Seeber went on to have a long and productive career at IBM.[4]

In terms of the subsequent history of the computer, the most important member of the Aiken team was Grace Hopper. Endowed with a brilliant intellect, she quickly learned the art of programming when she joined the Harvard-Navy project, and her knowledge of mathematics was a tremendous asset. One day, she later recalled, Aiken ordered her to produce the Manual of Operation for Mark I. Of course, Aiken supplied drafts of the text and carefully revised whatever she wrote, but in a real sense Grace Hopper was the principal author. When four articles were prepared for *Electrical Engineering* in 1946, it was again Grace who was called upon to do the actual writing. This time, however, her name appeared along with Aiken's as co-author. In later years she called Mark I her "favorite computer."

When Grace Hopper left the Cambridge computer project, she became a staff member of the new company of Eckert and Mauchly. She is credited with a major role in the introduction of the compiler. Later she was a primary author of the computer language COBOL. After transferring from the Naval Reserve to the regular Navy, she rose to the rank of admiral. Over many years, she built up a devoted following of friends and admirers, lecturing frequently on computers in general and on the distant past, recalling with fondness the days when she was introduced to the subject by Howard Aiken and the Mark I.

In the early days of the Navy operation of Mark I, according to Robert Campbell, the organization of the staff was "rather informal." Each member had a variety of jobs and helped with whatever problems

---

4. In a report written for the reunion celebrating the 25th anniversary of his graduation from Harvard, Seeber pointed with pride to his having had a "major part in all phases of conception, invention, design, constructing, and testing (and naming!) of the IBM selective sequence electronic calculator." Furthermore, he "had charge of its operation until it was torn down in 1952 to make way for a bigger and better 'brain.'" As a "co-inventor of this computer," Seeber boasted, he "had several patents issued, one of which contains claims fundamental to the entire computer field."

arose. Aiken, then "very much a hands-on manager," participated "directly in much of the work." Campbell particularly remembers Aiken's wife, Agnes, coming in occasionally to help out with the punching and checking of sequence tapes. Campbell and Bloch were particularly close to Commander Aiken (as members of his staff always called him during the war years).

One of the most important assignments given to the new machine was to solve certain problems concerning implosion. On 29 March 1944, while Aiken was still at Yorktown, John von Neumann, Chief of the Applied Mathematics Panel of the National Defense Research Committee, wrote to tell E. L. Chaffee that he had discussed with Warren Weaver "the possibility of carrying out some gas dynamical calculations in which various branches of the NDRC and the Services are interested." "We felt," von Neumann wrote, "that the possibility of making these calculations on your new device was an exceedingly tempting one." He would be able to come to Cambridge to discuss this in a couple of weeks.

On 22 March, Weaver had informed Chaffee that von Neumann was "considering undertaking certain important calculations, directly connected with the war, which might possibly be most favorably carried out on your new device," and that von Neumann would be getting in touch with Chaffee directly.[5]

In due course, it was agreed that von Neumann's problem would be run on Mark I. He arrived in early August 1944, accompanied by the mathematicians Valentine Bargmann and Charles Loewner, both of Brown University. Dick Bloch was assigned by Aiken to be primary programmer for the project, but he was given no indication of the purpose of the calculations or the context in which they had arisen. He was merely supplied with the conditions of the problem, which centered on a nonlinear partial differential equation of the second order that described "a spherically symmetric flow of a compressible fluid in the presence of a spherical detonation wave proceeding inward from the surface of the sphere toward its center." It was clear to Bloch from the start that standard techniques "using successive approximations" would not yield accurate and reliable results, and that difference equations would have to be used. "Accordingly," Bloch writes in *Makin' Numbers*, "$y = F(x,t)$ was to be determined numerically by creating a fine mesh in the two-dimensional $x$-$t$ plane across which a numerical

---

5. Note that both von Neumann and Weaver, lacking a generic noun to describe the new machine, called it "your new device."

solution for y and certain other variables important to the physical problem would be effected—mesh point by mesh point—within the pertinent domain." Organizing the mathematical methods of solution was quite time-consuming, and the subsequent programming was equally laborious. On the first run, the mesh was found to be too "coarse" and a finer one was introduced. Although von Neumann was responsible for the general plan for treating the problem, the day-to-day "dirty work of laying out the difference equations" was done by Bargmann and Loewner.

During the programming and running of the problem, von Neumann appeared in the Comp Lab only occasionally. According to Bloch, he showed no interest in the programming process and little interest in the nature and design of the machine.[6] He wanted to know only when the machine was actually running his problem, so that he could be on hand to watch the results being churned out—in particular, to find out whether the mesh that had been chosen was adequate to the purposes for which the calculation was being made. As Bloch recalls, von Neumann would "review the temperature and pressure output in order to determine whether the original differential equations and the subsequent difference equations were portraying this thing properly." Von Neumann "had his own way of knowing what was proper behavior." Some "rough calculations" he had made enabled him to know whether the results of a run were "good enough." On this basis, he would ask that the "fineness of the mesh be increased by a factor of 5 or 10."

Although the presence of von Neumann and two mathematicians made it obvious that the implosion calculations were of major importance to the war effort, no one in the Comp Lab was informed of the ultimate goal. Not until about a year later, when Hiroshima was bombed, did Aiken, Bloch, and other members of the Mark I staff become aware of the purpose of the implosion calculations.

During the days that von Neumann spent in Cambridge, his relations with Aiken were purely formal. Their meetings were brief and infrequent. This was, however, the beginning of a long friendship.[7]

6. While observing the implosion problem being programmed and run, von Neumann admitted to Bloch "that the computer area was pretty much a new problem to him."

7. In 1946, Aiken and von Neumann both served on the National Research Council's Committee on High Speed Calculating Machines. (See appendix F.) Later, when von Neumann found himself unable to accept Aiken's invitation to attend the second Harvard Symposium on Large-Scale Digital Calculating

Mathematicians and physicists in the greater Boston area got an opportunity to hear von Neumann lecture when Garrett Birkhoff[8] of Harvard's Mathematics Department arranged for him to give a special colloquium for the Physics Department. I had the opportunity to be present. Speaking clearly and distinctly in a loud voice, von Neumann darted from one end of the platform to the other, covering the blackboard with diagrams and equations. His subject was the use of multi-valued logic in quantum mechanics. Few members of the audience had both the necessary background in quantum mechanics and the requisite understanding of mathematical logic to follow the argument in full detail, but all could appreciate von Neumann's dynamic intellect and his enthusiasm.

Much later, long after Aiken was dead, it was revealed that calculations similar to those done at Harvard were being done simultaneously at Los Alamos in a kind of contest. In 1982, Nicholas Metropolis and E. C. Nelson published an essay on the computing facilities at Los Alamos in which the details of this competition were made public. While von Neumann, Bargmann, and Loewner were working with Bloch in Cambridge, the Los Alamos group was attacking the problem with modified IBM punched-card machines. Von Neumann apparently kept the Los Alamos group informed of the progress being made in Cambridge. The Los Alamos group completed its work in a much shorter time than the Cambridge group. However, "the punched-card machine operation computed values to six decimal places, whereas the Mark I computed values to eighteen decimal digits." Additionally, Mark I "integrated the partial differential equation at a much smaller interval size [or smaller mesh] and so . . . achieved far greater precision."[9]

In view of its use in solving the implosion problem, and the other applications described in this chapter, there can be no doubt that Mark I was indeed used to address significant problems related to World War II. But Mark I evidently did not perform all the war-

---

Machinery, he wrote (on 28 July 1949) not only to express regret but also to offer Aiken "heartiest congratulations" on the completion of Mark III. "I need not tell you," he wrote, "that Mark III and its operation interests me exceedingly."

8. Garrett Birkhoff, son of the above-mentioned George Birkhoff, was a distinguished mathematician in his own right.

9. Von Neumann did not bring any further problems to Aiken and Mark I.

related work it has sometimes been said to have performed for the Navy. For example, *Time,* in its issue of 23 January 1950, reported that Mark I had been used to evaluate "an electrically operated cannon that the Nazis were known to be building." According to this story, "Bessie" (the name *Time* used for Mark I) "chewed into a snarl of equations" and "proved that the weapon was entirely impractical." As a result, "the US relaxed while the Germans, who had no Bessie, went on wasting enormous effort on an impossible task." I asked both Dick Bloch and Bob Campbell whether such an assignment had been given to Mark I; both replied that they had never heard of it.[10]

In a 1960 talk titled "Computers—Past and Future," given at the Navy's David Taylor Model Basin in Washington, Aiken himself referred to certain wartime activities of the Harvard Comp Lab, of which he said he would "mention only a few." First he discussed the "extensive calculations in the development of radar for the Radiation Laboratory at MIT." Next he referred to the "investigation of metallurgical statistics" and "the calculations that . . . helped solve blast furnace difficulties that were causing considerable trouble at the time." The last of Aiken's examples was "the calculation and tabulation of a comprehensive set of tables, dealing with the vertical and horizontal components of the earth's magnetic field." These tables, he said, were "of tremendous importance in magnetic mine warfare." He did not mention the work done in design of telephoto lenses for the Army Air Force, nor did he mention the implosion calculations for the atom bomb, but perhaps that work was still classified.

10. Mary Aiken recalls that Howard was very proud of his activity in working out where to place mines along the Atlantic coast, an assignment that involved exploration in submarines. This was an appropriate asssignment for a specialist in naval mine warfare, especially one skilled—as Aiken was—in electomagnetism.

# 18

## The Mystery of the Number 23

One of the puzzling aspects of the design of Mark I is the provision for calculating to 23 places or digits (the 24th being used to indicate whether the number was positive or negative). The alleged purpose in endowing the machine with the capacity to compute to 23 digits, according to Aiken's early associates James Baker and Robert Campbell, was to enable it to perform calculations in celestial mechanics—in particular, to recalculate the orbits of the planets.

Grace Hopper told me that very often the 23 digits were more troublesome than helpful. For example, suppose a calculation were being made in which there were only seven significant figures. (The traditional rule is that actual calculations be carried to two or three more than the number of useful digits until the very end.) In performing such a calculation, Mark I would take all calculations out to 23 digits, 14 more than could possibly be useful—a tremendous waste of machine time. To bypass this feature of Mark I and enable it to perform certain calculations in a reasonable time, Grace Hopper and others would disable the extra digits and so reduce the actual machine operation to a smaller number of digits.

Herbert Grosch, in a critique of a published report of mine concerning the above-mentioned reason for endowing the machine with the capacity for calculating to 23 digits, expressed doubt that Aiken ever intended to compute orbits. He was certain that Aiken did not know any celestial mechanics. A trained astronomer, Grosch was aware that no problem in celestial mechanics would require 23-digit calculations.[1]

1. At Columbia University's Watson Computation Laboratory, Grosch had worked with Wallace Eckert on some of the most difficult calculations in the area of celestial mechanics: calculations connected with the motion of the moon.

I don't know how much Aiken actually knew of celestial mechanics, nor have I been able to find any informant who could enlighten me concerning Aiken's actual understanding of the problems of planetary orbit theory. Neither his undergraduate education at Wisconsin nor his training as a graduate student at Harvard included any courses on celestial mechanics. On the face of it, the number 23 seems absurdly extravagant for the computation of planetary orbits, and Grosch was quite right in doubting that this was a sound reason for giving the machine the capacity to compute to 23 digits. At the time, planetary observations were good to at most eight significant figures. Thus, even extravagant use of the doubling rule would require no more than sixteen digits.

The philosopher of science Ernest Nagel once remarked that history of science interested him because it turned up events and sequences of events that he would never have imagined on purely logical or intuitive grounds. In 1992, with this in mind, I sought help from Jim Baker, the celebrated designer of telescopes and lens systems. Baker seemed a particularly important witness not only because he had been trained in astronomy but also because (as was noted in chapter 17) the second problem assigned to Mark I had concerned a lens that Baker was designing. Furthermore, after the war Baker had brought a purely astronomical problem to Aiken. When I asked Baker why the machine had the capacity to calculate to 23 digits, he recalled instantly that orbital computations had been among the problems Aiken had proposed to solve with the new machine. Without a moment's hesitation, Baker said that he understood that Mark I had been designed with the capacity to compute to 23 figures specifically for the purpose of calculating planetary orbits.

For corroboration I turned to Bob Campbell, who had been closely associated with the design and initial operation of the machine. When I asked him if he remembered the reason for the choice of 23 digits, he said that Aiken had planned to recompute planetary orbits. Furthermore, in his historical account of Mark I in *Makin' Numbers* Campbell says:

> The reason for the 23 decimal digit precision is not generally known. I once queried Aiken on this point, and he replied that it had been planned to utilize the calculator to recompute the orbits of all the major planets in the solar system, and that a precision of 23 digits was judged to be necessary for this task. As it turned out, however, the machine was never used for this purpose.

There is other evidence to indicate that Aiken was interested in orbital calculations. For example, in the original proposal he submitted

Aiken as a boy. From the collection of Mary Aiken; reproduced with permission.

Aiken around the time of his graduation from high school. From the collection of Mary Aiken; reproduced with permission.

Aiken as a young engineer. From the collection of Mary Aiken; reproduced with permission.

Commander Howard Aiken, USNR, in summer uniform, 1945.

Robert Hawkins at work assembling Mark I, March 1944.

The IBM technicians who assembled Mark I at Harvard.

Mark I being assembled in the Battery Room of the Research Laboratory of Physics, March 1944.

Mark I in the Computation Laboratory, 1948.

A view of Mark I in 1944, showing (at extreme left) the constant registers and the storage registers.

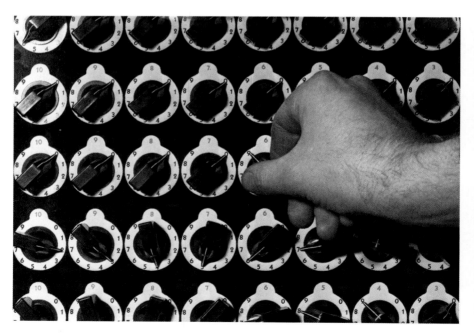

A close-up of the constant registers being set by hand.

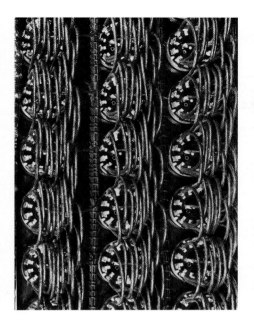 

Left: A rear view of the constant registers, showing the wiring connection to each position. Right: The jumble of wire connections constituting the nerve system of Mark I.

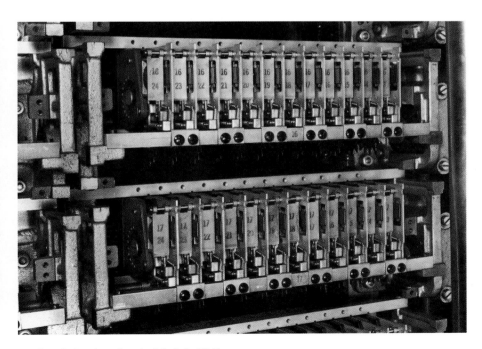

Banks of plug-in relays in Mark I, 1945.

Punched tapes used with Mark I.

Punched tape and reader, Mark I, 1944. Taken by A. J. Dionne of the Cruft Laboratory Photographic Department, Harvard University.

James Conant, an unidentified Naval officer, Benjamin Durfee, Clair Lake, Howard Aiken, Frank Hamilton, and Thomas Watson posed in front of the ASCC/Mark I, August 1944.

Dedication ceremonies for the ASCC/Mark I, Faculty Room, University Hall, Harvard University, August 1944. Among the men in the front row are Thomas Watson, James Conant, Leverett Saltonstall, and Howard Aiken.

The staff of Mark I in 1944. Seated in the second row, from left to right, are Richard Bloch, Hubert Arnold, Howard Aiken, Grace Hopper, and Robert Campbell. Robert Hawkins stands at the extreme left of the back row, David Wheatland at the extreme right.

Aiken in conversation with Ronald King in front of Mark I, 1944.

James Baker (extreme left) discussing a problem with Campbell and Aiken, 1944.

Richard Bloch, Howard Aiken, and Robert Hawkins, 1945.

Assembling Mark II.

Wiring Mark II, 1946.

Wassily Leontief and Howard Aiken in front of Mark II, 1947. Taken by Walter R. Fleischer of the Harvard News Office.

Mark II, early 1947.

Mark II, 1947.

The second story being added to the Howard Hathaway Aiken Computation Laboratory.

Mark III, 1949.

Tape mechanisms for Mark III.

Howard Aiken and two colleagues observing the operation of the "instructional tape preparation table" for Mark III.

Vacuum tubes, Mark III, 1948.

Cables, Mark III, 1949.

Cabling connections for Mark III, 1951.

Artzybasheff's rendition of Mark III for the cover of the 23 January 1950 issue of *Time*.

H. P. Babbage (great-grandson of Charles Babbage) and Howard Aiken in 1947, contemplating the calculating wheels donated to Harvard by Henry Prevost Babbage in 1886.

Mark IV, 1952.

Mark I (left) and Mark IV (right), assembled in the Comp Lab in the mid 1950s.

Reino Martin and his silver model of Mark I, 1961.

THE INSTITUTE FOR ADVANCED STUDY
SCHOOL OF MATHEMATICS
PRINCETON, NEW JERSEY

March 29, 1944

Dear Professor Chaffee:

    Last week I discussed with Dr. Warren Weaver, Chief of the Applied Mathematics Panel, N.D.R.C., the possibility of carrying out some gas dynamical calculations in which various branches of the N.D.R.C. and the Services are interested. We felt that the possibility of making these calculations on your new device was an exceedingly tempting one, and Dr. Weaver promised to get in touch with you on this matter. I have now learned from him that you have obligingly taken his proposal into consideration and suggested that I get in touch with you.

    I appreciate this opportunity exceedingly, and would like to discuss the matter in detail with you at your convenience. I regret that I shall not be able to come to Cambridge this week or next; but may I ask you which days in the week after next (April 10-15) would suit you?

    I am

                            Sincerely yours,

                            John von Neumann

Professor E. L. Chaffee
Electrical Engineering Department
Harvard University
Cambridge, Massachusetts
JvN:GB

John von Neumann's 1944 letter to E. L. Chaffee referring to the possibility of using Mark I for "some gas dynamical calculations."

The Howard Hathaway Aiken Computation Laboratory.

Howard Aiken celebrating his birthday.

Peter Calingaert, Howard Aiken, Frederick Brooks Jr., and William Wright on the occasion of Aiken's visit to the Department of Computer Science at the University of North Carolina, 1 March 1971. Courtesy University of North Carolina Photo Lab.

to IBM, the uses Aiken envisaged for the new machine included "astronomy."

Frank Hamilton's historical memorandum mentions that problems in astronomy were among the proposed assignments for the new machine. In those days, the major set of problems in astronomy requiring large-scale computation came from orbital calculations. At that time astronomy was one of the two primary areas of large-scale calculating, the other being the production of actuarial tables. In astronomy, a major computing activity was in the production of ephemerides—tables of positions of the planets and the moon for the Nautical Almanac. In the United States, this work was done at the Naval Observatory in Washington.

In his IBM oral-history interview, Ben Durfee recalled that Aiken had brought two problems to Endicott as samples of the assignments the machine would be given. One had involved bridge building; the other had been a problem in astronomy. I don't know whether Durfee's recollections on this point were accurate. Perhaps he was thinking of problems Campbell proposed after Aiken was called to active duty in the Navy. In *Makin' Numbers* Campbell writes that in autumn 1943 he "selected and programmed five test problems to demonstrate the capabilities of the machine, which was completely assembled and operational at Endicott." Campbell continues:

Actually, the "problems" were more in the nature of illustrative algorithms or subroutines. They were chiefly intended to exercise all of the built-in arithmetic and functional capabilities. The five problems were given the following titles:

an orbital calculation (with parameters for the planet Neptune).

current in the load of a transmission line

focal length of a plano-convex lens

catenary shape of a suspension bridge cable (with approximate parameters for the Williamsburg Bridge in New York City)

a "math problem" (this was actually the calculation of the log of the gamma function using Stirling's formula).

Two of these problems are the same as those Durfee remembered. It is far too many years after the event to know whether Campbell invented these problems entirely on his own or whether they were influenced by discussions with Aiken.

On the basis of this evidence, I believe that there can be little doubt that Aiken's references to using the machine for astronomy should be taken to mean calculating orbits, most likely those of the planets. I have found no basis for a belief that such problems would require calculations to 23 digits. Evidence from two independent and reliable witnesses, however, is not to be brushed aside easily. Both Baker and Campbell recall that Aiken said he intended to make orbital calculations and that this was the reason for having 23 digits.

The capacity to perform calculations to 23 places made the machine more complex (by requiring more working parts) and slower than it would have been if it had been built to calculate to only 16 or only 10 digits. Furthermore, the extra digits obviously added significantly to the machine's cost and its construction time. As is evident from the Durfee interview, the IBM engineers eliminated certain features (such as having more than one multiplier) in the interest of cost cutting. Surely, Aiken must have had a good reason for wanting 23 digits. Yet the only reason remembered by his associates was the desire to calculate orbits.

There is one further bit of evidence relating to the calculation of orbits: a report of a conversation or interview with Elbert ("Bert") Little, a longtime friend of Aiken's from his days as a graduate student. According to this report (found in the Aiken files in the Harvard University Archives), Little "recalled" that there had been a conversation "in which Professor Aiken asked Dr. Harlow Shapley (and perhaps other astronomers) how many digits were needed to plot the courses of the asteroids." In reply, "Dr. Shapley told Professor Aiken that over fifteen-digit accuracy was necessary, and Dr. Little believes that this information partially determined the nature of Mark I." Of course, this concerns orbits of asteroids rather than planets, and the number of digits mentioned is 15 rather than 23. Furthermore, at that time the Harvard astronomer Fred Whipple was computing the paths of asteroids, and it seems likely that Aiken would have gone to him rather than to Shapley for information on the paths of asteroids.

Bob Campbell remembers only Aiken's stating that he wanted to recompute the planetary orbits; he did not remember that Aiken ever spoke of any particular such computations—and, as he was quick to point out, the machine was never used for any such computations. Perhaps Aiken was only demonstrating his awareness of the importance of orbital calculations in astronomy.

In the days when the new machine was being planned, Aiken was in constant communication with Harlow Shapley, the director of the

Harvard College Observatory, who was certainly cognizant of the computational needs of astronomy. He had lectured on the use of machine calculation in astronomy. It was just such a lecture, as we have seen, that attracted the interest of Theodore Brown, astronomer and professor of statistics at Harvard's Graduate School of Business Administration, who had arranged Aiken's initial contact with IBM. Aiken also was associated with Theodore Sterne, a theoretical astronomer interested in many types of problems involving calculation. Another Harvard astronomer who had many discussions with Aiken, as Jim Baker recalled during my discussions with him, was Leland Cunningham, then a graduate student supporting himself by working for the Boston Edison Company. Cunningham's specialty was the computation of orbits, and he spent much of his later life doing orbital calculations by computer. Cunningham, according to Baker, would have been aware of the need for "a lot of extra places to allow for the round-offs." Baker noted that orbital calculations require a very large number of operations, and the number of round-off numbers needed goes up rapidly with the number of operations performed. "This is the reason I understood," Baker concluded, that Aiken had made provision for 23 digits.

There is some evidence to support Baker's recollections in Aiken's original draft of a talk for the dedication of Mark I. In this draft, though not in the actual presentation, Aiken thanked a number of Harvard colleagues: "Professors Chaffee, Mimno, Shapley, Brown and Mr. Cunningham, all of whom encouraged the development in the early days when the theoretical background of the machine was under consideration." The special mention of Cunningham proves that Aiken had been consulting with him about astronomical calculations. Since Cunningham's special interest was orbital calculation, it is reasonable to suppose that Aiken had been discussing that subject with him. It was, no doubt, at Aiken's suggestion that Cunningham was invited to the dedication.

Baker gave me some additional information concerning Aiken and orbit calculations. Around 1938 or 1939, when he was deep into the design of a "sequence controlled calculator," Aiken had several conferences with Fred Whipple (then teaching orbit theory in the astronomy department) and Cunningham. Baker said that Whipple and Cunningham, "perhaps with Shapley and perhaps with [Donald] Menzel as well," told Aiken about orbit problems, stressing that the computations "would have to go on for a long time" and that "the round-off accumulation would be a drawback to the orbital accuracy" after "perhaps six months." Baker suggests that "Howard went on to put together the safe margin."

While reminiscing about these events, Baker said that the information about Whipple and Cunningham came "directly from Aiken himself." Around the time when he went "over to the Computer Lab" ("It must have been in 1944, during wartime work"), Baker talked with Aiken about the number of digits. "One of my first questions," he said, was "why we should have so many significant figures." Aiken told him "about the conference" with Whipple and Cunningham and how the idea of a large number of significant figures "got started through orbital calculations." Baker then said that the information concerning the meetings between Whipple and Cunningham was not derived from his personal experience, but rather from "what Howard told me."

When I asked Campbell whether he remembered ever using 23 digits, he replied that this had been done during the computing of the Bessel functions. Richard Bloch had found the extra places necessary because of the normalizing procedure. In fact, during some of these calculations Bloch needed 46 places and found he could get them by "ganging up" the counters. At one time, Campbell said, there was the hope that it would be feasible to compute the lower orders and then use the results, by a recurrence technique, to obtain the higher ones. This did not prove to be the case, however.

When Aiken lectured at the Moore School in 1946, he described a kind of doubling that had been used to calculate to 46 digits in the process of computing tables of Bessel functions. It should be noted, however, that Aiken's next machine, Mark II, had provision for only ten digits and used a floating decimal point.

Mark I was planned and built as a fixed-decimal-point machine. Aiken discussed some of the consequences of this in the penultimate paragraph of his original prospectus. Having introduced the topic of "Suggested Accuracy," he pointed out that in an example given earlier in his presentation he had used ten "significant figures." He then asked how many "number positions" would be needed on "most of the computing components" if "all numbers were to be given to this accuracy" (ten digits). His answer was that "it would be necessary to provide 23 number positions," since this degree of accuracy would require "10 [digits] to the left of the decimal point, 12 to the right, and one for plus or minus." "Of the twelve to the right," he explained, "two would be guard places and thrown away."

Jim Baker alerted me to the fact that the advanced electromechanical desk calculators of the time (some made by Marchant and some by Monroe) had a feature that was equivalent to calculating with a large number of digits. Such machines were used in the late 1930s and later at the Harvard College Observatory. Although calculating to 23 digits seems excessive today, in those days the capacity to do so was not uncommon in advanced desk calculators. As Baker reminded me, these machines had in the center a mixing set of dials, which on the Marchant gave provision for 20 digits. "An upper dial, which was primarily for divisions or multiplicands and things of that sort, had eleven places, and above the main mixing group," Baker recalled, "were ten." The principle of operation "was that the center, which had the twenty columns, was a mixing system, which, whatever the calculations of the other two (in dividing or multiplying), contained the answer." Furthermore, "the decimal point could be placed anywhere, because there was no floating point." "So," Baker noted, "it may very well be [that] the desk calculator was that way because people usually could use their own judgment about moving decimal points back and forth, as I often did." In setting the decimal point, users would "regroup the significant numbers by a transport of maybe one, two, three or four decimal places at a time." Also, there was "[a] tendency in computing to group the numbers or digits, at least on sight, or when writing them down on paper, in sets of three, maybe four, at a time, however large the number was."

Aiken was familiar with the desk calculators of that day and had used them extensively. "The ordinary desk calculator [of those days] had at least twenty places in the mixing portion," Baker noted, "and so that may be what Aiken had in mind. He may have wanted to end up with ten significant figures, which would be adequate for orbit calculation. He may have had the idea of preserving ten significant figures somehow, on a non-floating point scheme."

In a personal communication commenting on this topic, Maurice Wilkes stressed that provision for 23 decimal digits does not imply that some problems require "an accuracy of 23 significant digits." "In some problems," he continues, "numbers vary widely in magnitude as the calculation proceeds and, if the registers are too short, it is difficult to prevent significant digits being lost either by overflow or underflow." This is a problem "met with in any computer without floating-point hardware."

It may be concluded that Aiken gave two very different reasons for having 23 numerical positions. As Baker and Campbell reported independently, he spoke of a desire to compute planetary orbits, which he presumably believed would require that number of places. He never made any such calculations, and had he done so he would by far have exceeded the limits of accuracy of observation. Indeed, as Herbert Grosch asserted, there is no evidence that Aiken knew anything about celestial mechanics or the computation of orbits. The second reason for having 23 places, set forth by Aiken in the prospectus, is that Mark I was originally conceived as a machine computing to ten places of accuracy on each side of the decimal point. This differs from making computations to 23 significant figures from start to finish. Yet in actual practice Mark I regularly computed to 23 places, no matter where the decimal point was placed.

# 19

## Tables of Bessel Functions

Most accounts of Mark I disparage its use to produce tables of Bessel functions. Often it is declared that Howard Aiken was addicted to making tables and "had a thing about" Bessel functions. Indeed, Aiken specified in paragraph 2 of the original proposal to IBM that the proposed machine should "be able to supply and utilize" on demand the values of certain "transcendental functions," such as trigonometric functions and "elliptic, Bessel, and probability functions." And the Harvard Computation Laboratory eventually used Mark I to produce twelve huge volumes of tables of Bessel functions, plus a volume of the related Hankel functions[1]—an activity that strikes present-day computer scientists as odd, since it is now customary to use special programs or built-in algorithms to compute values of functions as they are needed.

The term "Bessel functions" refers to a large class of mathematical functions that appear often in practical problems of science and engineering. They are named after Friedrich Wilhelm Bessel (1784–1846), a Prussian astronomer also known for his discovery of a way to measure the distance from Earth to a star. Although in 1824 Bessel devoted a special publication to the functions named after him, marking the formal beginning of the general use of these functions, some special instances of what we know today as Bessel functions had already been featured in the writings of Euler, Bernoulli, Lagrange, and Fourier.

In technical language, Bessel functions are solutions to a large class of differential equations of a form known as "Bessel's equation":

$$z^2 f''(z) + z f'(z) + (z^2 - n^2) f(z) = 0.$$

---

1. A special set of Bessel functions—the "third kind"—named after the mathematician Hermann Hankel (1819–1873).

Because this is a differential equation of the second degree, there are two "families" of solutions. Classified as Bessel functions of the first and the second "kind," these are further subdivided by "order" (indicated by a subscript J—for example, $J_3$ stands for a Bessel function of the third order).

Bessel functions appear in a wide variety of scientific problems in the life sciences and the physical sciences—problems having to do with such things as the vibration of metal plates, FM radio signals, and the structure of DNA. It is the job of the mathematician addressing such problems to decide which of the functions is applicable and then to find the approximation that most closely fits the conditions within some designated range of the variables. As late as the 1960s, applied mathematicians consulted printed tables of functions rather than making laborious individual calculations.

One such set of tables, published in 1937, was produced by the Committee for the Calculation of Mathematical Tables of the British Association for the Advancement of Science. In July 1948, "responsibility for the work on mathematical tabulation" was transferred from the British Association to the Royal Society. So great was the demand for the tables that they were reissued in 1950 and again in 1958. During the 1950s, the Royal Society also published a number of other volumes of tables, including *A Short Table* for another class of Bessel functions (1952), *Bessel Functions and Formulae* (1953), and *A Short Table for Bessel Functions of Integer Orders and Large Arguments* (1954). During the 1960s, I made regular use of the *Handbook of Mathematical Tables,* a 1964 supplement to the *Handbook of Chemistry and Physics* published by the Chemical Rubber Company. Many applied mathematicians still make use of the 1046-page *Handbook of Mathematical Functions, with Formulas, Graphs, and Mathematical Table,* edited by Milton Abromovitz and Irene Stegun and issued in 1964 by the National Bureau of Standards.

It was once common practice to interrupt a computation in order to locate the appropriate volume of tables, look up the value needed for the problem at hand, then return to the computer and enter that value. A primary assignment of early computers or super-calculators such as the ASCC/Mark I, ENIAC,[2] and even EDSAC was to produce numerical tables. Later, computers made it relatively easy to compute values of some functions as they were needed.

---

2. ENIAC was commissioned specifically to compute ballistic tables.

Richard Bloch's recollection of how the project of computing tables of Bessel functions came into being suggests that the idea of using Mark I to prepare such tables did not originate with Aiken or with any of the members of his Cambridge group. In the spring of 1944, when Bloch first visited Cambridge to see the machine in operation, Robert Campbell recorded in the logbook that Bloch had come from the Naval Research Laboratory and that he was interested in Bessel functions. One of the first programs Bloch wrote for Mark I after he joined Aiken's staff in Cambridge was a program for producing tables of Bessel functions.

Before Dick Bloch joined Aiken's staff, he worked in the radio division of the Naval Research Laboratory. Bloch has reported that at one meeting of his research group the mathematician Bernard Salzberg complained that he "needed tables of Bessel functions in order to solve certain problems of radar antenna patterns." When Bloch joined Aiken's staff, he brought Salzbergs's request to Aiken's attention. Bloch observes that Salzberg probably was thinking of tables to six places, but that Aiken "naturally decided to produce tables to four times as many places." Very likely, Bloch comments, Salzberg "didn't ask for $J_0$ through $J_{100}$," which were eventually computed on Mark I and published in the Annals of the Computation Laboratory of Harvard University. As this story makes plain, the Navy's desire for such tables was a major reason for its support of the machine's operation.

The tables of Hankel functions were published in 1945 in the first volume of the Annals to be printed and distributed.[3] The formal title was *Tables of Modified Hankel Functions of Order One-Third and of Their Derivatives*. As the preface explains, these tables were produced in response to a specific request from MIT's Radiation Laboratory: "The need for solutions of Stokes' differential equation in the complex plane arose during the war in connection with a problem of Wendell H. Furry, at that time a member of the staff of the Radiation Laboratory." It was "requested that the Bureau of Ships of the United States Navy assign to the Computation Project the task of computing the present tables." The preface says that the control tapes were "coded at the end

---

3. I remember the general applause when Aiken came to a meeting of the Harvard Physics Department—at which Wendell Furry was present—to exhibit the first volume (which was designated volume 2 so that the Manual of Operation for the Automatic Sequence Controlled Calculator could be called volume 1).

of August 1944 by Lt(jg) Grace Murray Hopper, USNR, after consultation with Professor [Wendell] Furry," and that "unfortunately, Lt(jg) Hopper was taken seriously ill during the computation," whereupon "the computation was transferred to Ens. Ruth A. Brendel, USNR." The preface also records that throughout "the running of the problem and the preparation of the tables Mr. Robert L. Hawkins and Mr. David P. Wheatland gave invaluable and constant assistance."[4]

Although Mark I was designed as an all-purpose machine, it was especially well adapted for producing tables in that it could be programmed to provide solutions with regularly increasing increments of the independent variable. Accordingly, it continued to crank out such tables, probably to more orders and more significant figures than any ordinary mathematician would ever have needed.

The real innovation in tablemaking represented by Mark I is evident from the contrast between the way Mark I produced printed tables and the traditional ways of doing so, as described in L. J. Comrie's introductory essay to the 1937 British tables. Comrie reports proudly that the work had been done primarily "on Hollerith, National, and Burroughs machines." (Some had been done on a Brunsviga model 20.) All values had been "computed to two decimals more than are given in these tables," so that "the error of any tabulated value should not exceed" a predetermined quantity. For the benefit of those engaged in "special work," the "original values, with two extra decimals," were "available on application to the Secretary of the British Association."

Mark I's tremendous advantage in tablemaking is nowhere more evident than in the transition from computed output to published volume. Mark I's printouts of computed tables, produced by a pair of electric typewriters, were simply photographed and printed, whereas the British publications required laborious proofreading and checking. In his introduction, Comrie explained at some length how the "proofs were read against the printer's copy" (which, as has been mentioned,

---

4. Although Aiken stressed the importance of direct reproduction of the tabular output, the first published volume of tables was actually produced in a quite different manner. As was explained in the preface, a result of external circumstances was that the computation of the tables of Hankel functions had to be done "as a base load problem employing the calculator at such times as other problems would permit." Because of this "mode of operation," it was necessary "to retype the tables rather than use the results as printed by the calculator."

had been produced by mechanical calculation). The proofs were also read "against the original calculations, containing two or three decimals more than are given here." Comrie paid tribute to the "compositors and readers at the Cambridge University Press" that "not a single compositor's error was found" in the "280 pages of tables" ("or just under a million figures"). And this was not the end of the checking process: the proofs "were read, as far as possible, against published tables." And, just to make certain that no last-minute errors had been introduced during the printing process, which was based on printing "from movable type," there was "a further precaution": the "final printed sheets were checked by differencing all values with a National machine to the fourth difference." "In so doing," Comrie continued, "the printed second differences were also set on the machine, and the usual differencing process modified in order to compare the printed second differences with those derived from the function values." Comrie boasted, in conclusion, that he had "every reason to believe that the tables are completely free from error."

As further evidence of the advantage of using computers to produce tables of functions, consider *Tables of the Reciprocal of the Gamma Function for Complex Argument,* a volume of about fifty double-sided folio pages issued in 1950 by the Computation Centre of the University of Toronto. These tables were produced by John Pearson Stanley of the University of Toronto and Maurice Wilkes of Cambridge University. This work was apparently mimeographed (or produced by some offset process from typed pages), and the pages were stapled together. A preface explained that the "entire computation" had been "performed on the electronic digital computer EDSAC at the Mathematical Laboratory, Cambridge, England." Values were "quoted to six decimal places, with a maximum error of a unit in the last decimal place." In order to "minimize rounding-off errors, all computations were carried to ten decimal places." After the original calculations were completed, the "table was checked by differencing in both directions on a National accounting machine." Boasting that this was "the first tabulation of the complex gamma function which is at all extensive," the editors informed users that the table had been "checked against all previous tabulations." Nevertheless, the compilers found it necessary to issue a page of errata.

Shortly after the publication of the Manual of Operations for the ASCC/Mark I and the first volume of tables produced at Harvard, L. J. Comrie published in *Nature* (then, as now, the world's foremost general scientific journal) the review I have discussed in chapter 8. Comrie especially noted the way in which the output of the new machine had been transformed into print—the means of "conveying computed results to many users." Reproduction "by photolithography" of tables "typed by an electromagnetic typewriter," Comrie admitted, "eliminates many fruitful sources of error and much drudgery, transcription, comparison and proof-reading." The importance of this statement becomes apparent when one takes note of Comrie's description (quoted above) of how his tables were proofread. (Oddly, although Comrie's general theme is that the ASCC/Mark I is an embodiment of Charles Babbage's concepts, he does not mention that Babbage proposed a way to convert the tabular output of his machines directly into type, without copying, typesetting, and proofreading.)

It is not easy to justify the time, labor, and expense of producing Aiken's volumes of tables of Bessel functions.[5] Recall that there were twelve large volumes, containing tables for orders 0 through 100, and that values were given to 23 digits. The continued reprinting of Comrie's much simpler tables seems to indicate that applied mathematicians and engineers did not, in ordinary practice, need or use Aiken's tables. I have tried without success to find to what extent these volumes of tables were actually used. Most applied mathematicians seem to think that Aiken's tables constituted a kind of overload. Only an exceptional problem would need tables of more than six or eight digits.

It is hard to escape the feeling that, once he had started computing tables, Aiken just couldn't stop. And after the basic program for computing these tables had been worked out, the subsequent programming was relatively simple and straightforward. Like many other military projects, the computation of tables (initiated by a request from the Navy) was kept going without any direct inquiry into the practicality of the product. Furthermore, since the Navy was continuing to support his Computation Laboratory, Aiken may have believed that

5. Volumes 3–14 of the Annals of the Computation Laboratory were devoted to Bessel functions. Later volumes were devoted to tables of generalized sine and cosine-integral functions (volumes 18 and 19), tables of inverse hyperbolic functions (volume 20), tables of generalized exponential integral functions (volume 21), tables of the function $(\sin f)/f$ and its first 11 derivatives (volume 22), and tables of the error function and its first 20 derivatives (volume 23). Of the first 35 volumes in the series, 19 consisted of tables of functions. The last such volume was *Tables of the Functions arc sin z* (volume 40), published in 1964.

he was bound to keep showing his sponsors tangible evidence of his activity. The correspondence files in the Harvard University Archives show that while Mark I was "cranking out" tables Aiken refused requests from outside the armed forces and the university to use the machine. The continuing production of these tables without reference to any actual use for them suggests that Aiken was isolated from some of the main streams of computer activity—an isolation that became more and more apparent with the passage of time.

In a final judgment about the making of tables, we must be careful not to consider the worth of tables of functions from the point of view of today, when computers use special programs or built-in algorithms to compute needed values of functions rapidly and easily as needed. I have cited a number of tables published in the immediate postwar years that indicate that the use of Mark I to produce tables was not in itself extraordinary even if the degree and scope of Aiken's tables may seem excessive.

The activities of the National Bureau of Standards, in publishing many volumes of tables, provides yet another index of the importance of mathematical tables during the 1940s, the 1950s, and the 1960s. Some of these were said to have been "prepared by the Mathematical Tables Project, conducted under the sponsorship of the National Bureau of Standards." For example, the tables of "arc sin $x$" (published in 1945) was said to have been "begun under the auspices of the Work Projects Administration of the city of New York and completed with the support of the Office of Scientific Research and Development." The Harvard Library contains a total of 76 volumes of numerical tables in this series, including several devoted to Bessel functions, the last of which was published as late as 1967.[6] Clearly such tables did not at once cease to be useful with the advent of the computer.

6. An important study of this topic has been made by David Alan Grier: "The Math Tables Project of the WPA: The Reluctant Start of the Computer Era," *Annals of the History of Computing* 20 (1998): 33–50. Grier is in the process of writing an extensive history of table making.

# 20
# Aiken's Harvard Program in Computer Science

One of Howard Aiken's most important pioneering activities—the one that may have been his most significant contribution to computer science—was the establishment of a graduate program in computer science at Harvard for the academic year 1947–48. This program, based on a series of courses that offered opportunities for hands-on experience using Mark I,[1] led to the Master of Science degree.

The term "computer science" did not come into general usage until some time after the inauguration of the Harvard program. Yet, as Robert Ashenhurst notes, "computer science" was practiced at Harvard "in an academic setting a good ten years before departments with that name started appearing on the university scene."

Although Aiken's Harvard program was the first to offer both an M.S. and (in 1949) a Ph.D., it was not the first program of instruction in computer science. In the academic year 1946–47, courses in computer science were being taught at Columbia by Wallace Eckert[2] and Herbert Grosch.

In 1947, when Harvard began its program in computer science, Harvard was one of the few universities with a large-scale computing machine available. The Moore School at the University of Pennsylvania had ENIAC; however, there was no accompanying academic program, and ENIAC was moved to the Army Ordnance Laboratory at

---

1. Later, students were able to work with Mark IV and Harvard's UNIVAC and to pursue the Ph.D. degree.

2. Eckert had returned to Columbia after leaving in 1940 to become director of the Nautical Almanac at the Naval Observatory in Washington. On 1 March 1945, while remaining a professor at Columbia, he had become the director of IBM's Pure Science Department. A 1951 Columbia brochure about the computer program lists him as "Director of Pure Science" (his IBM title) and "Professor of Celestial Mechanics" (his Columbia title).

Aberdeen, Maryland. Furthermore, J. Presper Eckert and John Mauchly left the Moore School to found their company.

On the last day of the Harvard Symposium on Large-Scale Digital Calculating Machinery (10 January 1947), Aiken announced his plans to establish a program to develop computer scientists:

> It has become increasingly clear that we must start a training program. We must remember that our universities are primarily institutions for the building of men and not for the building of machines, and we must offer courses of instruction in this field. I feel that one of the most important contributions that the Staff of this Laboratory can now make is, with the coming of the next school year, to offer courses of instruction in applied mathematics with a strong flavor of computing machinery. I sincerely hope that, with the permission of the Faculty of this University, we shall find and develop methods of furthering this purpose.

Aiken had long been aware that the future of computers would depend on the availability of trained mathematicians who would have the skill to convert scientific and engineering problems into a form in which they could be fed into a machine, and who would be the programmers of the machines.

Aiken was notoriously pessimistic about the availability of mathematicians capable of programming large-scale computing machines. Indeed, he did not see how there could ever be enough mathematicians available to program them. It must be remembered that Mark I, at the peak of its operation, required the services of four fully trained mathematicians, all working overtime. At that rate, 25 machines would require more than 100 mathematicians—a number that by far exceeded the pool of available talent. In the years just before World War II, only a very small number of doctorates were awarded in mathematics.

In the first full academic year after the war, 1946–47, Harvard's catalogue listed no courses on computers, although "Associate Professor Aiken" then directed research (Applied Science 20t) on "numerical analysis and design of calculating instruments." In the next year, though, Aiken got his training program underway—characteristically, on his own terms. Beginning with the fall of 1947, the Department of Engineering Sciences and Applied Physics offered a one-year program leading to a Master of Science degree in applied mathematics "with special reference to computing machinery." The success of the program, enlarged in 1948 to accommodate the Ph.D. degree, can be seen in the fact that 19 M.S. degrees and eight Ph.D. degrees[3] were awarded between 1948 and 1954.

Jack Palmer was a student in the master's program in 1947–48. He had returned to Harvard to finish his undergraduate work in mathematics after completing his military service in World War II. He heard about the new program and decided that Aiken and his associates were pioneering an interesting and promising field, and so he enrolled in the new computer science[4] program and obtained his master's degree. When I asked Palmer about his thoughts of a future career at the time he entered Aiken's graduate program, he remembered anticipating a need for programmers.[5]

At first the Navy, which had underwritten the costs of running Mark I during World War II, underwrote the new Harvard program through the recently established Office of Naval Research. It was agreed that the Navy could nominate 15 candidates (who would have to meet Harvard's admission standards) and Harvard another 15. However, the Air Force took over the financial support of the program, having signed a contract under which Aiken would train Air Force personnel, pursue research in many aspects of computers, and eventually construct a large-scale computer. Bert Little served as liaison between the Air Force and Aiken's operation at Harvard.

In those days, at Harvard, in contrast with many other universities, the master's degree required no thesis but only the completion of eight half-courses. The minimum requirement in Aiken's program was one half-course in statistics, two half-courses in numerical analysis, and other half-courses, including one (Applied Science 218, taught by Aiken) on "the organization of large-scale calculating machinery." The subject matter of Applied Science 218 was officially described as "the

---

3. For administrative reasons, all the doctorates save the first two were officially awarded in Applied Mathematics.

4. The academic subject we know as computer science did yet have a name. The new machines were not yet generally known as "computers"; they were called "calculators" or "computing machines." The subject of Aiken's two Harvard symposia (1947 and 1949) was "large-scale computing machinery." Aiken, John von Neumann, and others tended to refer to "computing machines"; the word "computer" still designated a person rather than a machine. The term "computer science" seems to have come into usage in the late 1950s.

5. In fact, Palmer subsequently had a distinguished career as a programmer at IBM and eventually became a member of IBM's team of historians.

design and construction of large-scale digital calculating machinery and the application of such machines to the solution of scientific problems." There was also an important half-course (Applied Science 248, taught by Harry Mimno) on "electronic control and calculating circuits"; it remained a popular and important component of the program for many years. Aside from the announced subject matter, "electronic timing circuits and associated apparatus applicable to large-scale digital machinery," Mimno addressed the design of circuits for electronic telemetry.

The course on numerical analysis (Applied Science 211–212), taught by Kaiser Kunz, stressed interpolation, numerical differentiation and integration, and numerical solutions to ordinary and partial differential equations and integral equations, plus an "introduction to large-scale digital calculating machines" and the "preparation of mathematical problems for large-scale calculators." Each student was expected to "prepare and run a problem." The statistics half-course (Applied Science 215) included "applications of large-scale computing machinery in theoretical and applied statistics."

Graduate students could also pursue a course of assigned reading and research with Aiken on "numerical analysis and design of calculating instruments." In the first two years, Aiken later reported, 76 students had been enrolled in the program and 14 students had earned the M.S. degree. In 1948, Herbert Francis Mitchell Jr. was awarded a Ph.D., the first in this area.

By the academic year 1949–50, the program was in full swing. The catalogue of courses listed two half-courses on numerical analysis (Applied Science 211–212) to be taught by Assistant Professor Wei Dong Woo; the statistics course was to be taught by Assistant Professor David Middleton. Woo was also listed as giving a reading and research course on "automatic computing devices." In 1950–51, Woo was on a medical leave of absence; his course on numerical analysis was taught by Miles Hayes (who received his Ph.D. under Aiken in 1950), while Mimno's course on "electronic control and calculating circuits" was given in his absence by Peter Elias, a member of the Harvard Society of Fellows who later had a distinguished career at MIT. There were three undergraduates and 14 graduate students (reduced to eight in the spring semester) in the course in numerical analysis, one senior and 14 graduate students in the statistics course, and one undergraduate and 14 graduate students in Aiken's course on large-scale calculating machinery. Four graduate students and one undergraduate were enrolled for reading and research under Aiken's direction.

In 1951–52, the description of Aiken's course was revised to read as follows:

> This course deals with the principles of the design and construction of large-scale digital calculating machinery and the application of such machines to the solution of scientific problems. Emphasis is given the mathematical synthesis of electronic calculating and control circuits, using algebraic methods developed in the course. Electromechanical techniques are exemplified by the Mark I computer, relay techniques by the Mark II computer, and electronic, magnetic, and dry rectifier techniques by the Mark III and Mark IV computers.

The next year, this description was shortened to read:

> Mathematical methods for the synthesis of relay, rectifier, and electronic switching circuits with applications to calculating machinery.

The course was now listed as Applied Mathematics 218 (no longer Applied Science 218). In 1951–52 Aiken's students numbered two sophomores, one junior, one senior, one special student, three graduate students, and one Radcliffe undergraduate; in 1952–53 there were two juniors, two seniors, and ten graduate students.

Jack Palmer took Aiken's Applied Science 212 course in the spring term of 1950. His lecture notes show that the students were required to spend two days a week in the lab. The first six sessions were devoted to a problem to be programmed on Mark I. Subsequent sessions dealt with matrices (three meetings), bivariate interpolation, cubature, and integral equations (three meetings), partial differential equations associated with Laplace's equation (three meetings), a vibrating drumhead (three meetings), and a vibrating string (two meetings). In the final two sessions, students, working in pairs, were scheduled to "operate Mark I for approximately four hours." Aiken prepared a list of 36 problems, from which each pair of students was to choose one—subject to the "instructor's approval." Aiken wisely advised the students to select problems for which "the mathematical analysis" was already "familiar to them." They would be allowed only 24 hours for the problem, of which six would be "allotted to the actual operation of the calculator" with a "short preliminary period of instruction in the operation of the machine." The assignment consisted of the following steps: each "student team will perform the necessary mathematical analysis, organize the machine solution, design a complete checking system, code the control tapes, prepare the input data, compile the operating instructions and plugging diagrams, prepare re-run instructions of what to do in case of a check stop, and prepare and operate the calculator to solve the problem."

As I read the 1950 guide prepared for Aiken's students, I could not help but be aware of the near-identity of many of these step-by-step instructions with those I had been given when I wrote a student program of my own a dozen or more years later under the tutelage of Ken Powell at IBM. A major difference was that my instructions did not include "a complete checking system" or a set of "re-run instructions of what to do in case of a check stop." Furthermore, the later instructions stressed the significance of "debugging" the program. Aiken's instructions emphasized his constant insistence on accuracy and reliability.

To facilitate the student's operation of Mark I, a mimeographed "students' handbook" was prepared. Comprising about 100 pages, this book dealt with all the practical steps needed for the successful programming and running of problems. According to an introductory statement, the handbook's purpose was to give students the information they would need to operate Mark I. "The point of view is strictly mathematical," the handbook declared, "No electrical or mechanical information is given." References were made to the 1946 Manual of Operation, to the set of three articles in *Electrical Engineering,* and to the privately produced Handbook of Coding Procedures for the Automatic Sequence Controlled Calculator (March 1948), with this warning:

The above references should be treated with caution as they are obsolete in some details, particularly the C codes there given. Changes are being made in Mark I all the time so even this book may be previous or obsolete. It is our aim to have it correct as of April, 1950.

Distinguishing between "functions performed by machine" and "functions performed by mathematician," the students' handbook gave a neat encapsulated statement of Aiken's philosophy. The "functions performed by the mathematician" ("assuming he is also the machine operator"), fall into three categories:

1) Mathematic analysis applicable to any digital computer . . .
2) Paperwork peculiar to Mark I, including coding, preparing plugging diagrams and operating instructions . . .
3) Physical preparation and operation of Mark I.

The "general purpose of mathematical analysis" was

to find a solution of the given problem in the form of a sequence of simple algebraic equations, each one of which involves only a single one of the four arithmetic operations (+−×÷) and in which the letters represent finite, discrete, terminated numbers, that is, numbers which can be placed in a 24-column counter.

Since the students were required to have taken the first half-course in numerical analysis, in which they would have been introduced to Mark I and would have been given the student's handbook, they were already acquainted with Aiken's philosophy of machines by the time they took his course.

Jack Palmer recalls that in 1949–50 Aiken was always available to the students, as were a miscellaneous assortment of instructors, graduate students, and staff members. He remembers with particular fondness Aiken's course in "organization of large-scale calculating machinery" (Applied Science 218), given primarily for graduate students. The "reading period" assignment was for the whole class (consisting of eight students) to produce jointly a report of a "preliminary design of a low cost computing machine." At the time, $35,000 was considered a low cost for a computer. It was assumed that the construction would be done in a university laboratory. The computer was to be designed for use by "a consulting engineer on engineering problems." The students were to suppose that a mathematician would be "in charge of the machine"—that is, available to reduce the problems to a form suitable for solution by machine and to write the programs. The goal was to provide "sufficient accuracy for engineering problems." The machine was to occupy no more than 8 square feet of floor space. It would "operate at a speed perhaps of the order of fifty times as fast as the Mark I calculator." In a day when machines often bore acronyms as names—including such fanciful ones as MADAM, JOHNIAC, BINAC, ILLIAC, SWAC—the students spent a good bit of time devising a name that would yield an amusing acronym. They finally chose Simple Arithmetic Digital Sequential Automatic Calculator (SADSAC).

The concluding paragraph of the students' handbook sets forth Aiken's specifications of the limits of the assignment. Portions of that concise statement of his philosophy of computer design, quoted earlier, will be repeated here since it relates to the students' assignment. "The 'design' of a low-cost computing machine," it is said, "is understood to consist in the outlining of its general specifications and the carrying through of a rational determination of its functions." In this context, "design," it is noted, "does not include the actual engineering design of component units." As we have seen, this statement summarizes Aiken's philosophy of the role to be played by universities in computer design and is an accurate statement of his role in the "design" of the ASCC/Mark I.

One aspect of Harry Mimno's course (Applied Science 248) on "electronic control and calculating circuits," given again in the spring of 1950, was "the design and operation of Eccles-Jordan trigger circuits." Thus the students were introduced to what was then agreed to be the basis of all electronic computing. The final exam included questions on general features of computers (such as the "block diagram of operation function table") and on specific hardware (including a "binary parallel adder"). Students were asked to compare and contrast "the early models of electronic computation devices (as represented by the ENIAK [sic] design) with current design practices" and to summarize "the features of the Mark III design, as presented in the reading-period reference."

Mimno's final exams often exhibited signs of his penchant for encouraging originality within rigid prescribed requirements. On the final exam for 1950, for example, Mimno allowed the students to substitute for any one of the assigned questions "a description and discussion of an original design, based upon the general class of electronic devices studied in the course."

The June 1950 final exam for Applied Science 218 asked the student to produce "a single line diagram for a large-scale digital computer" and to "discuss the major components." Another problem was to derive "mathematical expressions for the circuits of a serial binary adder for three quantities, obtain vacuum tube operators for the required circuits and sketch them." The students were expected to be familiar with binary-coded decimals, since they were asked to derive "vacuum tube operators for a doubling circuit, assuming that the decimal digits are coded in the 8, 4, 2, 1 system." The final question concerned "a drum coated with a thin layer of magnetic material," on which a series of binary digits could be "recorded as dipoles in a single chain channel," the digits 1 and 0 being represented by dipoles with oppositely directed orientations. The assignment was to draw a "symbolic diagram or a circuit" to transform these digits to a different channel of the same drum. The student was expected to give the wave shape for each point of the circuit and "a timing diagram of any clock pulses used in the operation."

In addition, the students in the master's degree program took courses in electronics and in mathematics. In applied mathematics they had to learn how to find roots of equations ("to slide rule accuracy") such as

$$\sin x = \frac{1}{\sqrt{x}}$$

and

$$x^4 - 3x^3 + 3x^2 - 3x + 1 = 0.$$

Students were expected to be able to derive "Simpson's 1/3 Rule," to deal with difference equations and the method of successive differences, to use the method of least squares, and to calculate definite integrals. Another course in applied mathematics introduced matrices, sets of four simultaneous linear equations of the first degree in four unknowns, the use of Laplace's equation, the eigenvalues of a vibrating plate, and integral equations. There can be no doubt that Aiken's students obtained a well-rounded education.

Garrett Birkhoff, a longtime colleague of Aiken's, recalls that Aiken was, despite his conservatism in some areas, "very forward looking." In particular, according to Birkhoff, Aiken "saw that scientific computing was all very well, but that we must not put all our undergraduate and masters' candidates into numerical analysis and scientific computing." Rather, Aiken believed that students should be encouraged to enter "data processing." The Committee on Data Processing (chaired by Albert Hartlein, Gordon McKay Professor of Civil Engineering) had among its members Howard Aiken (of course), Ted Brown of the Business School, the economist Wassily Leontief, Harry Mimno of Applied Physics, some other engineers and economists with the rank of associate professor and assistant professor, Kenneth Iverson (who had recently completed his doctorate with a thesis directed jointly by Aiken and Leontief), and Robert Minnick.

In addition to the masters' degree program, there was also a complement of graduate students and young instructors.[6] The roster of

6. An Wang, the founder of Wang Laboratories, became a member of Aiken's staff in 1948 upon completing the requirements for his Ph.D. For his recollections of his days at the Comp Lab, see his book *Lessons: An Autobiography* (Addison-Wesley, 1986).

Aiken's doctoral students includes a number of figures who later became prominent in the computer world. While Tropp and I were interviewing Aiken in the living room of his Fort Lauderdale home, he pointed with manifest pride to a shelf above the couch on which he sat. On the shelf was a set of bound theses of his Ph.D. students. The former students, Aiken said, were important contributors to computer science and the computer industry. It was obvious that he considered that to have had such students was one of his major claims to a significant place in history.

Aiken's favorite topic of instruction—the subject he considered to be at the heart of understanding computers—was switching theory, a major part of his principal course on computers, Applied Mathematics 218. The term "switching theory" is not in general use today as it was in the past. The *Encyclopedia of Computer Science and Engineering* (under "Education," on page 591) defines switching theory as "theoretical principles and mathematical techniques involved in the design of digital system logic." When Tropp and I interviewed Aiken in 1973, he asked me whether "the course on switching theory" was still taught at Harvard.

Aiken organized two symposia on the exploration of switching theory and its applications. The first of these provided the text for volumes 29 and 30 of the Annals of the Harvard Computation Laboratory, published in 1959. The second symposium arose, according to Aiken, in 1961, after Aiken and David Grinnell Willis "encountered a number of switching problems peculiar to space applications" and decided that "the switching problems of space-vehicle control" needed special attention. This project won the support of John Purcell Nash, then Director of Research (and soon to be Vice-President for Research and Engineering) at Lockheed. Willis was then the manager of Lockheed's Computer Research Department, and Aiken was serving Lockheed as a special consultant. The symposium, supported by the Air Force, was held on 27–28 February and 1 March 1962. Willis left Lockheed before the proceedings were edited for publication under the title *Switching Theory in Space Technology* (1963). In the preface, Aiken explained that before "the advent of large-scale digital computers, switching theory was concerned primarily with the theory of relay contact networks." The scope of switching theory, he continued, had been greatly broadened in recent years by "the invention of many new switches, including rectifiers, transistors, and magnetic cores, and the burgeoning growth of both variety and number of applications." There had been much research in this field, stimulated by these "new devices and applications." But, as is characteristic of "any new area," there was a paucity of interactions among investigators. Aiken expressed his hope that the symposium would not only inform researchers about what others were doing in this new area but also help to establish channels of communication among researchers in the field.

Commenting on the second symposium, Aiken told Tropp and me that he had been able to get people from "all over the world" to attend. (Some 25 papers were presented to an audience of more than 700, representing ten countries.) The goal was to "see if we couldn't get the space industry to make more sense in the equipment they were designing."

# 21

## Later Relations between Aiken and IBM

By the mid 1950s it was evident that Howard Aiken's program at Harvard was turning out well-trained computer scientists at both the master's and the doctoral level. IBM needed such individuals for its ever-expanding programs in the computer area. Thomas J. Watson Jr., who had taken his father's place at IBM, regretted that IBM's relations with Harvard University had been damaged by the rift between his father and Aiken and hoped that steps could be taken to establish friendly relations. With that goal in mind, the younger Watson lent his full support to Cuthbert Hurd's efforts to recruit computer scientists from the students in the Harvard program. This was a part of Hurd's plan to develop a staff of men with higher degrees in mathematics and the sciences who would become contacts between IBM and the rapidly growing market of the scientific and engineering communities.

Hurd, an able mathematician, was the second (or third) Ph.D. to be hired by IBM.[1] He had received a Ph.D. in mathematics from the University of Illinois in 1936, taught at the Coast Guard Academy from 1942 to 1945, and then served as a dean at Allegheny College. From 1945 to 1947 he had held a senior research position at the Oak Ridge National Laboratory, where he had his first experience with IBM machines. After going to work for IBM in 1949, he had risen quickly to become head of the corporation's recently created Department of Applied Science. A courteous and affable man with a strong character and a considerable knowledge of mathematics, he had hit it off with Aiken right away, and they were to remain friends for the remainder of Aiken's life.

1. Hurd had attended the second Symposium on Large-Scale Digital Calculating Machinery, held at Harvard in 1949.

Hurd's efforts to establish friendly relations between IBM and Aiken had the strong support of Vincent Learson, a Harvard alumnus who was a vice-president of IBM. In the summer of 1996, six months or so before Learson died, Hurd—hoping that Learson would consent to being interviewed by me and Robert Seidel of the Charles Babbage Institute—arranged for me to speak with him by telephone.

Learson, Hurd had told me, had been present at the dedication of the Mark I. When we had established telephone contact, I asked him whether he remembered much about that event. He replied that his recollections of it were somewhat vague. He supposed that "the old man" (T. J. Watson Sr.) had asked him to attend. (Learson was then a very junior member of IBM's staff, and perhaps the reason Watson asked him to be there was that he was apparently the only Harvard graduate on the IBM payroll. Rex Seeber, the second Harvard graduate to be employed by IBM, was still working in Aiken's laboratory at the time.) Following Hurd's suggestion, I kept the conversation brief. But before we ended, I asked Learson about an incident concerning him and Aiken that has been widely circulated, even appearing in print. I wanted to find out from Learson whether this reported event had actually happened.

In 1972, the story goes, Learson paid a visit to an exhibit called "A Computer Perspective," which consisted of a three-dimensional "history wall" embellished with photographs, documents, and artifacts. Designed by the Office of Charles and Ray Eames, this detailed presentation of computer history was housed in the lobby of IBM's old headquarters at 590 Madison Avenue in New York.[2] Learson spent considerable time studying the displays. As was reported to me (and as is recorded in the literature), when he got to the panel devoted to the contributions of Howard Aiken he paused, glared at Aiken's portrait, and muttered (loud enough for everyone present to hear) "That son of a bitch!" Learson laughed when I recounted this story. "I don't remember saying that," he said, "but I certainly would have." Learson's antipathy toward Aiken was not just the reaction of an IBMer taking the side of the company against an ungrateful outsider. Learson

2. One reason I was particularly interested in the story of Learson's visit is that in those days I was serving as IBM's principal historical consultant, and I had worked closely with the Eames Office on the historical background of the exhibit. I had first heard about the incident of Learson's visit from one of the IBM employees who conducted visitors through the exhibit.

had actually met Aiken, but only once. The encounter had ended disastrously.³

In an interview with Hurd, which I conducted in the summer of 1983, he told me that he and Learson had decided to arrange a special occasion that would bring Aiken to Endicott in order to meet Thomas J. Watson Jr. As a first gesture, they would show Aiken what IBM was doing in the general area of computers and calculating. As Hurd explained to me, the aim was to "get [Aiken] to feel better about us and get us to feel better about him."

Hurd and Learson arranged for Aiken to be driven by limousine from his home in Winchester to Logan Airport in Boston, where one of IBM's private planes was waiting. Hurd was at the airport in Endicott, waiting to greet Aiken and to accompany him to the laboratory, where Watson Jr. and Learson were waiting.

Hurd remembered that the four men had a "pleasant but fairly stiff luncheon." At first, everything seemed to be going very well. Watson said that he was very happy that Aiken had come to Endicott, and that he would be pleased to take Aiken through the laboratory. Aiken expressed genuine interest in finding out what IBM was doing in the computer area. Watson then told Aiken that before they actually entered the laboratory, he would be expected to sign the non-disclosure agreement that all IBM employees, consultants, and visitors were expected to sign. The purpose of the agreement was to protect IBM's proprietary interests in novelties not yet released or patented. When Aiken refused to sign, Watson remarked that this was merely a routine procedure. Aiken solemnly declared that he had never in his life signed such an agreement, and he never would.⁴ Everything he did and everything that was done at Harvard was in the open, in the public domain. He wasn't concerned with secrets and patents. He was interested only in knowledge and teaching. He "never received confidential information."

It is unlikely that there were any top-secret developments that would have had to be kept from Aiken's eyes even if he did not sign a

---

3. Learson was quite friendly toward Harvard later; indeed, he served on the Visiting Committee of the Division of Applied Science. He was also involved in a major plan to obtain funding for computer science at the university.

4. Neither Aiken nor Watson nor Hurd nor Learson seems to have been aware that Aiken had signed just such a non-disclosure agreement with IBM when the ASCC/Mark I was being planned and designed.

non-disclosure agreement. If there had been some, the tour could have been arranged so that Aiken would not have seen them. The IBM group surely should have known that Aiken was sensitive on many issues and that he was easily set off. They seem to have rather unthinkingly provoked his stormy response, and they do not seem to have been prepared for his out-and-out refusal. My guess is that Watson considered the non-disclosure agreement to be so routine and ordinary that he was taken aback by Aiken's blatant refusal.

After Aiken made his declaration of principles, there was a moment of silence. Then Aiken stood up and asked Hurd to arrange for him to be taken home. Hurd called the airport to alert the pilot, then escorted Aiken to the waiting limousine, which took him back to the airport.

For the rest of his life, Aiken rarely if ever had a kind word to say about IBM.[5] Yet Hurd told me that Aiken's personal relations with Watson Sr. and Watson Jr., and with IBM's "administration generally," never "interfered with Harvard-IBM relations." Furthermore, said Hurd, Aiken "was always complimentary of the IBM engineers who had worked on the [ASCC/Mark I] project."

5. He did want to know what was going on at IBM, and so he bought one share of stock so as to receive the annual report and all the other literature about new products and research that the corporation sent to its stockholders.

# 22
## Aiken at Harvard, 1945–1961

Aiken's activities during the late 1940s and the 1950s show how the computer was gaining recognition as an aid to research in many different areas. The variety of work done in the Comp Lab during these years also reveals the important place of Aiken and his Harvard computing unit in the new world of computers. Much of Aiken's energy during the 1940s and the early 1950s was spent in the design and construction of three more giant machines. He also was among the first to recognize that computers would have important uses in the world of business.

When Aiken returned to civilian life after World War II, he was an associate professor. In 1946 he became a professor of applied mathematics, a post he held until his retirement in 1961. He became a member of the newly created Department of Engineering Sciences and Applied Physics, which included the old Cruft group plus a host of new faculty members whose appointments had been made possible by a very large bequest from Gordon McKay. In 1949 this department was merged with parts of the old Graduate School of Engineering to become the Division of Engineering Sciences as part of the Harvard Faculty of Arts of Sciences.

The year 1946 saw the construction of a new and separate building to house the Comp Lab, with space for Mark I and the staff and students. This building was conceived as a first step in building a Harvard "Science City," a program that never was realized. The new building was set off by itself, apart from and not connected to the nearby Jefferson Physical Laboratory and Physics Research Laboratory, the Cruft Laboratory, and Pierce Hall (the center for engineering and electrical technology). This physical separation seemed to symbolize Aiken's aloofness from the rest of the faculty concerned with pure and applied mathematics and physics. Later, bridges were built between the Comp Lab and neighboring Pierce Hall.

Funds for the new building came in part from IBM's gift of $100,000 at the time of the dedication of the ASCC/Mark I in 1944 and from the $4000-per-month rental fee paid to Harvard by the Navy for the use of Mark I. The new building, known for years as the Computation Laboratory and later designated the Howard Hathaway Aiken Computation Laboratory, contained faculty offices, student workrooms, lecture halls and classrooms, a machine shop, a 60-by-60-foot space to house computers, and an imposing atrium lit by large two-story windows. An exhibit case displayed some of the books of Charles Babbage, Harvard's set of Babbage computing wheels, and some other early computational devices. Visitors could watch the computers at work through a clear glass wall. After Mark I was decommissioned, its place was taken by a UNIVAC, Harvard's first commercially produced computer.

The Comp Lab soon proved inadequate to house the rapidly growing program, and a second story was added. By 1997 the building no longer met the needs of computer science at Harvard. The huge, atrium-like central space, designed to hold Mark I and Mark IV side by side, was no longer well suited to the computers of the 1990s. The building was razed to make way for a much larger and more up-to-date center for computer science.

Aiken's reputation as a leader in the new domain of computers was aggrandized by his success in organizing two international symposia on "large-scale computing machines," held in Cambridge under the sponsorship of the Harvard Computation Laboratory and the Navy. The published proceedings of the first symposium, held in 1947, show that almost every major figure in the new world of computers was there. J. Presper Eckert and John von Neumann are conspicuous by their absence, but each was represented by a close associate (John Mauchly in the case of Eckert, and Herman Goldstine in the case of von Neumann). Not surprising, however, is the absence of Konrad Zuse; the importance of his innovations was not yet fully recognized. The second symposium, held in 1949, had an even larger and more impressive international attendance. One of the great virtues of these symposia was that they brought together the computer pioneers who were advancing the new art and science but who did not yet have any established avenues of contact and exchange of ideas. There were 336 in attendance at the first symposium, 701 at the second.

In his introductory remarks at the second Harvard Symposium on Large-Scale Digital Calculating Machinery, in September 1949, Aiken noted that "an ever-increasing number of industries [were] interested in constructing computing machines outside the universities." "At our laboratory," he announced, "we have decided not to undertake the construction of any more large-scale computing machines with the exception of one, which we hope to build for our own use and keep at Harvard." He stressed that the true role of universities ought now to be "to offer well-rounded programs in numerical methods and the application of computer machinery" and "to train future operators of these machines." The rest of Aiken's career at Harvard was devoted to just such a training program and to finding new applications for the computers designed and built by industry. These new areas of application arose from various needs, some coming from the military and from other branches of government, others from industry (notably the utility industries), and yet others from research and instruction by Harvard students and faculty.

At the time of this announcement, Aiken had been responsible for two large-scale computing machines, Mark II and Mark III, designed and constructed for the Navy. Mark II was in service at the Naval Proving Ground in Dahlgren, and Mark III was under construction. Aiken was then in the preliminary stages of planning his last machine: Mark IV, commissioned by the Air Force. Mark II was built at the request of the Navy during the last year of World War II. Because of the Navy's desire to have this machine as quickly as possible, the design did not incorporate any radical new concepts; the result was a sort of improved Mark I, using relays of a new type. Mark II was delivered to the Navy in 1948. Mark III also ended up at Dahlgren.

Mark II, like Mark I, was a relay machine, but it had a floating rather than a fixed decimal point. It was a decimal machine that computed to ten places. Mark III, which used both vacuum tubes and solid-state devices, was Aiken's first electronic machine.

One of the innovative features of Mark III was a coding machine that Grace Hopper described as a unit "the likes of which was never built for any other computer." This device was designed to minimize error in the writing of programs. One source of such error was a fault in copying standard instructions into a program. With Mark III, a programmer could merely push several buttons and the lines of code for a series of equations would be automatically entered into the program.

Mark IV, an all-electronic machine, was built under a contract with the Air Force. It remained at Harvard, where it had a long and useful life. It served not only the Air Force but also outside users, Harvard faculty members, and a generation of students. Among its innovative features was the use of selenium diodes (later replaced by germanium diodes).

None of Aiken's four machines embodied any of the features of the new concept of the stored program. As the record plainly shows, Aiken's machines had no real influence on the main line of the rapidly developing design of computers. By the time Mark II was put into operation, a new age had dawned, of which the visible sign was ENIAC. EDVAC and EDSAC announced the stored program and in a very real sense defined the modern computer. These steps were so gigantic and revolutionary that, in retrospect, they almost dwarf the magnificent achievement of Mark I. Although Mark III and Mark IV did useful work during their lifetimes, they were far from being state-of-the-art machines in the sense that Mark I was in 1944. Indeed, the very fact that these later machines were used for important work is a sign of the need for computing power in the late 1940s and the early 1950s.

During Aiken's years at Harvard, the Comp Lab was a busy place, a center of activity where students, technicians, and faculty enjoyed a friendly atmosphere and learned as much from contact with one another as from formal instruction in aspects of computer design, operation, and application. The pleasant atmosphere of the Comp Lab contrasted with the often stormy relations between Aiken and a succession of Deans of the Faculty. The difficulties between Aiken and the deans arose not only from Aiken's demanding and somewhat aggressive attitude but also from the fact that the deans, who obviously did not recognize that Aiken was drawing Harvard into a leading role in the new computer age, constantly refused to give him real financial support.

One part of Aiken's agreement with the Air Force involved the design, construction, and operation of Mark IV. The other part specified that a number of Air Force officers would come to Harvard each year to learn about computers and their operation. One condition that Aiken set was that some time would be reserved for the growing computer needs of the Harvard faculty. Even so, Mark IV could not satisfy the ever-increasing computing needs of faculty research.

By the late 1950s, Mark I was more a historical monument than a useful working tool. Accordingly, it was dismantled and divided into three parts. One part was sent to the Smithsonian Institution, another to IBM to be added to the corporation's historical collections; the remaining part was put on display at Harvard.[1] To meet the research needs of the faculty and to provide services for others, the Comp Lab acquired its first stored-program computer. The archives record that in November 1956 the "installation of a UNIVAC System in the Computation Laboratory was completed." The system comprised a UNIVAC Central Computer and "peripheral equipment consisting of a Unityper (or input preparation device), a Card-to-Tape Converter, a Tape-to-Card Converter (80 column card) and a High Speed Printer."

Since one of the reasons for obtaining the UNIVAC was to serve the computing needs of members of the Harvard faculty, notice was given that staff members of the Comp Lab were available for "consultation on programming problems." Additionally, the staff of the Comp Lab gave "two series of lectures on elementary programming" for the benefit of faculty members and others interested in using the new machine. The National Science Foundation had made a grant to support faculty projects that had no funds of their own to pay for computer time. The response of the faculty was overwhelming. The archives contain detailed information concerning applications for use of the new computer. A Faculty Committee on Computational Facilities was established to review proposals for using the computer facilities and to establish orders of priority for the various proposals. Wassily Leontief of the Economics Department was the chairman; the other members were Aiken, E. B. Wilson of the Chemistry Department, Warren Semon (assistant director of the Comp Lab), and Kenneth Iverson.

In 1957, Robin Esch and Peter Calingaert prepared a manual to help inexperienced users do problems on the UNIVAC. Esch, then teaching Applied Mathematics 212, needed computer time, which was granted. The committee voted to make time available to qualified graduate students for work in connection with their dissertations. Aiken encouraged the members of the Harvard faculty and their graduate students to explore new uses of computers in various areas of research, and he and his staff were very helpful to those members of the Harvard community who brought their computational problems to the Comp Lab.

One of the first "non-scientist" faculty members to use Mark I was Wassily Leontief. Aiken's helpfulness to Leontief formed the basis of a long-term collegial relationship. In 1997, I had occasion to discuss with

1. Initially, the Harvard portion was displayed in the atrium-like lobby of the Comp Lab. Since 1998 it has been on view in the Harvard Science Center.

Leontief the details of his first encounter with Aiken and the new machine. I asked him about a statement that had been made to me by Betty Jennings, one of Aiken's former secretaries. She had said that Leontief had always been welcome as a visitor to the laboratory, and that Aiken had never quarreled with him. Leontief replied quickly and in the affirmative: "I don't remember having any quarrel with him." He also remarked that he was "probably one of the few" members of the Harvard faculty to be "users of the big machine."

Leontief is famous for his invention of "input-output" economics, which later earned him a Nobel Prize. Basically, Leontief's system of economic analysis consists of constructing huge two-dimensional tables showing what each industry receives from and what it sells to every other industry. In this way, Leontief was able to show in great detail how income flows through the whole economic system and to analyze how any changes are disseminated so as to affect the whole economic system. With greater refinement, Leontief's system of input-output analysis has been able to deal not only with whole industries but even with subsections of industries. An example, cited by Leontief's fellow economist John Kenneth Galbraith, is the effect of an increase in automobile production, which produces demands on the steel industry, which in turn result in demands on coal production and a host of other industries. Of enormous importance in Leontief's analysis of the US economy has been the absorption of specific resources by the military and the consequent flow of money.

In 1967, Owen Gingerich (a Harvard astronomer and historian of science) and I interviewed Leontief as part of our historical research into the early use of computers. Leontief told us he had used Mark I in his research. He explained that in developing input-output economics he had faced the enormous difficulty of having to find "solutions to systems of 15 or 20 [differential] equations." In the 1930s and the early 1940s, he explained, there was no way for him to produce his analysis other than "by hand on a small, small kind of desk calculator." Leontief described to us how he had tried to solve his problems by using a giant mechanical analog device invented and constructed by John Wilbur of MIT. The results had been disappointing, he said, and so he was overjoyed to learn of Aiken's machine. He recalled his delight when he found he could assign problems to Mark I and his satisfaction on finding that "the computer did it all."

During one of our telephone conversations in 1987, I asked Leontief what he remembered about his first contact with Aiken. Had he sought out Aiken, or had Aiken learned of his work and invited him to use the new big machine? We were talking about events of a half-century earlier, and Leontief wasn't certain; however, as he remembered, he found out about the "big machine" at the Computation Laboratory and then went to see Aiken to find out whether he could use the laboratory's facilities for his problems. Aiken apparently agreed at once that the differential equations and matrices in Leontief's research were suited to solution by machine and worthy of being assigned to Mark I.

Aiken was a real promoter, always on the lookout to find new fields of research for the new breed of giant machines. After Leontief had begun to work on Mark I, Aiken invited him to give a paper at the 1947 Symposium on Large-Scale Calculating Machines. Leontief's report on the work he had done on Mark I ("Computational Problems Arising in Connection with Economic Analysis of Interindustrial Relationships") was subsequently published in the proceedings. In the published report, Leontief showed how the central problem of input-output economics required the "solution of systems of linear equations" and the "computation of complete inverse matrices." Up to the "present time," he reported, "such computations have been performed on matrices of the tenth, twentieth, and fortieth degree." In the case of the fortieth degree, however, "no attempts have been made to invert the matrix, and the system of forty equations has been solved only for a few sets of the 'free' constants."

The final part of Leontief's report is noteworthy. He explains that both "the theoretical preliminary analysis and the collection of factual statistical information are progressing at such a pace that the economist will very soon be looking for means of solving very large systems of simultaneous linear differential equations." If "there had been any doubt whether the science of numerical computation would be able to come to his assistance," he concludes, "this symposium has certainly dispelled it by giving an unmistakenly affirmative answer." Leontief may be the first social scientist to have used a computer.[2]

Aiken and Leontief quickly established cordial relations. They arranged to jointly supervise the doctoral dissertation of Kenneth Iverson, later to become famous for his authorship of the language

2. During the 1973 interview, Aiken referred to some work that Mark I had done for an economist named Friedman. He was not sure who Friedman was but seemed to remember that he was a Chicago economist. When Tropp suggested that this may have been Milton Friedman, Aiken agreed. It was not clear, however, whether this work was done before or after Leontief's or for what subject Friedman needed the calculations.

APL. The very title of this dissertation indicates the combination of economics and computer science: "Machine Solutions of Linear Differential Equations: Applications to a Dynamic Economic Model." Before long, Leontief was on a number of committees related to the use of the Harvard computer.

The joint sponsorship of Iverson's doctoral research was but one example of Aiken's eagerness to establish interdepartmental collaboration. Another of Aiken's graduate students, Tony Oettinger, earned his doctorate jointly in computer science and linguistics. Oettinger, described by Aiken as "my heir," worked with Aiken in computer science and with Josiah Whatmough in linguistics, producing a doctoral dissertation that explored the use of a computer to translate Russian. The linguist Roman Jakobson was also involved in this project.

Another Harvard faculty member who came to Aiken with a computer-related problem was John Edsall, for many years a professor of biochemical science. Edsall had two separate encounters with Aiken and reported that his "personal experience with Aiken was, on the whole, certainly a very pleasant one." In the early 1950s, Edsall and Harold Scheraga (of Cornell University) had been studying a problem in "evaluating non-Newtonian viscosity" (the "variation of viscosity with velocity gradient") in "systems of ellipsoidal particles." At that time it was assumed that protein molecules could be considered "to be ellipsoids of revolution"; the goal of Aiken and Scheraga was to "calculate the dimensions of the ellipsoid." (Recalling this episode some 45 years later, Edsall remarked that we know now that this is "not a very good approximation.")

On 1 March 1952, Edsall—then a visiting Fulbright lecturer at the University of Cambridge—wrote to Aiken about his problem. Essentially, what was needed was to "calculate the dimensions of the ellipsoid" from the known viscosity of the system and the experimentally determined "amount of double refraction of the velocity gradient." Edsall informed Aiken that the equations awaiting solution were "mathematically very closely related to the problem of double refraction of flow which the Computation Laboratory succeeded in solving so thoroughly in 1949." This earlier work was reported in 1949 at the Second Symposium on Large-Scale Digital Calculating Machinery. The paper, "Double Refraction of Flow and the Dimensions of Large Asymmetric Molecules," appeared in the published proceedings under the names of three authors: Scheraga, Edsall, and J. Orten Gadd Jr. of the Harvard Computation Laboratory. (Gadd had been assigned by Aiken to program the solution and to run it on Mark I.) One passage in the report is of special interest:

> The total machine time consumed by this problem was slightly over two weeks. It is felt that here is a perfect example of the situation where a large-scale automatic calculator has a tremendous advantage over hand or desk computers, perhaps not so much in the matter of speed as in the problem of organization.

Aiken replied to Edsall's letter at once, expressing interest in Edsall's new problem and in "finding a means by which this work may be undertaken." Edsall's project would have been especially pleasing to Aiken because there were research funds available to pay the costs of computer time and the services of the programmer. But, as has been noted, Aiken had a grant from the National Science Foundation to cover such costs for faculty members who had no research grants. After returning to Harvard in the autumn 1952, Edsall visited Aiken, who received him graciously. Edsall told me: "I remember him looking over the equations in the papers we took to him, and saying 'Yes, this is just the sort of thing we can solve.'" After the problem was programmed and put to the computer, a full technical report was published in a professional journal.

I have mentioned on several occasions in this volume that Aiken either liked or disliked an individual from the start. In my own case, I felt we hit it off at once. He treated me like a fellow professional, a representative of a field of scholarship that he valued not merely for its own sake but also for its potential as a new area for computer application. I was more than a little apprehensive when I first went to see him in order to discuss a project of mine. After all, Aiken was a full professor with an international reputation in his field. He was widely known for being the "boss," the "commander," of his computer laboratory, while I was a lowly member of the junior faculty. I had telephoned to ask for an appointment without specifying in detail why I wanted to see him. My fears quickly disappeared when he at once got down to business.

The purpose of my visit was to explore with Aiken an aspect of the research I was then undertaking in collaboration with Alexandre Koyré. We were planning an edition of Isaac Newton's great book, the *Principia* (*Mathematical Principles of Natural Philosophy*), to be based on a collation of the three authorized printed editions and the original manuscript. I had come to see Aiken in order to explore with him the possibility of collating the variant texts by computer.

Aiken listened to my proposal, asked a few very pertinent questions, and then had me describe some of the features of the proposed edition. The moment I mentioned the problem of collating texts, Aiken "took off" and began to work out the details of a new application for the computer. He at once envisaged having the texts of the three editions entered on tape and then having the machine run through them and note any differences. "All we need is to find a good graduate student," he told me, "one who can work up this problem and devise a program for this purpose." "This," he continued, "could develop into a thesis subject." The only ingredient I would have to provide was the money to pay for it all. Aiken had already launched a large-scale program of textual analysis by computer: a concordance of the Bible, undertaken by John Ellison for the Nelson Publishing Company.

Before I left Aiken's office, he had worked out almost every detail of a plan. He would find an able graduate student who was willing to take on this project and to program it and see it through the stages of computer manipulation as a basis for his doctoral dissertation. He did some rapid calculations and worked out exactly how much financial support I would need to have in order to pay for the student's work and the machine time, plus the cost of having an operator enter the several texts into the computer and then proofread them. He even discussed in detail the qualifications that the student would have to have, how he would go about finding the student, and how much time it would take.

Before I could get a word in edgewise, Aiken had organized the whole project, explaining to me just how the computer would function in comparing the different texts, noting the places where they were similar and dissimilar. In retrospect, what I found most impressive about this experience was Aiken's joy in finding a new area of application for the computer and in working out the details of a new research project. Of course, I was also deeply moved by Aiken's expression of respect for my chosen professional discipline of history of science, something that was not shared by some of his fellow scientists. Above all, I was astonished and pleased that Aiken did not treat me as a junior or as a supplicant (both of which I, in fact, was), but as a colleague whose company he found significant because I had an interesting problem. He was not concerned with whether I was a professor, an associate professor, or a lowly assistant professor.

When Aiken began to sketch out how my collation could be done by computer, his enthusiasm was so infectious that I confess that I was thrilled by the prospect of inaugurating a new scholarly technique while preparing the new edition of Newton's masterpiece. In the event, the work of collation was done in the old-fashioned way, by hand, line by line. The reason is that when I sought the advice (and possible financial help) of Warren Weaver, one of Aiken's teachers at Wisconsin, who was then at the Rockefeller Foundation, Weaver strongly urged me to make the analysis of the texts by hand rather than by machine. The reason, he said, was that if I did the work myself, I would learn about the successive alterations made by Newton and be better able to understand their significance, whereas if the job were done by machine I would not have the same learning experience. In the end, Weaver proved to be a wise counselor, and I did not become a pioneer in machine comparison of texts.

Another area that attracted Aiken's attention was the use of computers in various aspects of business. This involved not only a course in the use of computers for business and the introduction of a program for would-be executives in the Harvard Business School but also an active program of working with businesses of various types in developing applications of computers.

Fred Brooks recalls how Aiken once chose "a topic [for him] to investigate," one about which Aiken held strong convictions. The theme Aiken wanted Brooks to examine was that "the design of special purpose machines for particular business applications would show substantial savings and improvement in cost performance because of the specialization of the application." Brooks was given the payroll application. "After a year of work," Brooks reports, "it became very evident to me that the payroll application did not in fact allow any particular gains over any other of the serial tape file maintenance applications—utility billing, insurance premium accounting." Brooks found that "the class of machines for which specialization was profitable was the serial tape file maintenance machines, not any particular subclass." In the end, "even though that went quite against the presuppositions with which [Aiken] had started, he respected that view and understood it, and I think adopted it."

Brooks's experience leads naturally to a very interesting aspect of Aiken's pioneering efforts: the use of computers in industry. During the 1973 interview, Aiken freely admitted that at first he had thought of machines only in the context of solving large-scale scientific and engineering problems. It had never occurred to him in the early days that such machines might have any use in business or in banking. He then saw their function to be to perform the calculations necessary for solving mathematical equations of the kind encountered in science and engineering and to produce mathematical tables. George Chase (the chief engineer at Monroe, to whom Aiken went to explore the possibility of building his machine) had envisaged from the start that computers would have an important role in the commercial domain. In this context it must be remembered that the giant early machines were chiefly one-of-a-kind, designed for specific purposes (chiefly engineering and applied science and government tasks). Furthermore, even when computers began to be manufactured, there was for a long time a separate product line for business and another for science and engineering. IBM did not produce a uniform set of computers that could be adapted equally for business and for science and engineering until System /360. After all, as Aiken explained during the 1973 interview, in business you are dealing with a large amount of input data and an equally large amount of output but very few operations, whereas in science and engineering the situation is just the reverse: the input and the output tend to be much simpler, the operations to be performed much more complex.

A talk given by Aiken in 1954 at the Harvard Graduate School of Business Administration reveals the strength of his belief about the impossibility of using computers for business purposes. "I do not exaggerate," he told his audience, "when I say that everyone who was in any way connected with the original development of large-scale computers had in mind only one thing—namely, the construction of machines for the solution of *scientific* problems." He then admitted: "No one was more surprised than I was when I found out that these machines were ultimately going to be instruments which could be used for control in business."

In 1954, the year when Aiken spoke about computers in business or industry, a survey showed that only about 5 percent of computers were being used in business applications. The only available program of instruction that did not concentrate on mathematical applications and the design of computers, it was noted, was the one scheduled to start in the next academic year at Harvard. Designed by Aiken, this proposed new venture in conjunction with the Harvard Business School was "noteworthy for its breadth, with courses in accounting . . . , in statistics, in data processing . . . as well as the more established courses in numerical analysis and in electronic control and calculating circuits." At this time, Aiken made a plea that young men be found who would be equally interested in the social sciences and computing devices: "We simply cannot stay within narrow boundaries." He stressed the "need for continuing the kind of research to stimulate each to think about the other's field."

One of Aiken's ways to advance the applications of the computer in business was to send an emissary to find out what the actual state of things was. Tony Oettinger reports that Kenneth Iverson was sent out to read gas meters and that he himself went to H. P. Hood and Sons, a large New England distributor of milk, to study their payroll problems. Iverson later worked on the computer billing project sponsored by the American Gas Association. When Oettinger completed his degree, in 1954, Aiken "sent" him to work for the Philadelphia National Bank—chosen, no doubt, because the bank's officers seemed sympathetic to new ideas and new ways. Oettinger says that almost at once "we started inventing things like account numbering schemes." Two major supporters of Aiken's Comp Lab at Harvard in the 1950s were the Bell Telephone Laboratories and the American Gas Association/Edison Electric Institute. Computerized billing by public utilities goes back to the Harvard Comp Lab and to the explorations made by Aiken for the utility companies.

Payment for this research was an important part of the support for the activities of the Comp Lab, although the amounts seem small by today's standards. The American Gas Institute and the Edison Electric Institute each paid $100,000 over a three-year period in the 1950s, the Bell Telephone Laboratories some $10,000 a year.

During the 1973 interview, Aiken described an aspect of his work with the utility industry in which his job had been, in part, to correct what he called "an operating mistake." In those days, he said, the "way public utility bills were made" was by "punching in" the "last month's meter reading" on a card. Then, after "meter readings were brought in for this month, you had to punch the meter readings for this month." You "then took these cards and put them through a calculating punch that took the difference and punched the difference between the two months of consumption on the same card." "You then sorted the cards by consumptions," Aiken explained. Because "you had pre-computed a table" that the machine could read, you knew "the bill for one kilowatt and two kilowatts and three kilowatts" and so on. "So you took all the one kilowatt cards and the bill for one kilowatt, and you put those in a duplicating punch and punched the bill; and you did this all the way up to a hundred thousand kilowatts and so on." He then told Tropp and me that when "programmable computers were available commercially to the utility industry, they initially programmed the machines to compute the electric bills in the same way." In other words, the utility companies treated the bills on their computers in exactly the same manner as when the bills had been sorted by hand. The only difference, Aiken noted, was that the utility companies "stored the table of the bill for $n$ kilowatts, and the sorting was done on tape." Even so, he agreed, "the programming procedure employed was exactly the same." In fact, "It took a little while to convince them there was another way." As he recalled this "amusing mistake," Aiken burst out laughing.

# 23
## Life in the Comp Lab

One of Aiken's favorite pupils, Robert Ashenhurst, recalls that for Aiken "theory and practice could [and must] interact and reinforce each other in many fields related to the construction and application of computing devices." Aiken's graduate students served as assistants in the courses and were encouraged to give courses of their own as a step up the ladder. Kenneth Iverson has recalled graduate study under Aiken as "like an apprenticeship" in which the student "learned the tools of the scholarship trade." Every topic was "used more as a focus for the development of skills such as clarity of thought and expression than as an end in itself." Once admitted to the program, a graduate student underwent a rite of "adoption into the fold." He was given a desk (or a share of a desk) among a group of other graduate students, the permanent staff, or visiting scholars, "most of whom were engaged in some aspect of the design and building of computers." A student was thus "made to feel part of a scholarly enterprise" and was provided, "often for the first time, with easy and intimate access to others more experienced in his chosen field."

Aiken stressed "the development of the ability to speak and write." Even before a student had finished his course work, he might be urged by Aiken "to participate in the work of the lab in some way, such as serving as a teaching assistant or taking summer or other part-time work supported by research contracts that he had negotiated." Aiken tended to give his graduate teaching assistants "considerable responsibility," even to the degree of "participating in the design of question sets as well as grading them, giving occasional lectures when the professor was out of town, or preparing and delivering the lectures for some specific part of the course—such as the introduction to computer programming."

Aiken was especially concerned with clarity of exposition. "Every thesis student," Iverson recalls, "knew and dreaded the fact that Aiken would carefully read and discuss successive chapters of the thesis, insisting upon clear exposition and offering both general and detailed advice." One effect of Aiken's emphasis on writing style was that "budding authors soon learned to call upon their fellow graduate students and other colleagues to review their work." Once a student had a chapter analyzed by Aiken, he would never submit another without the revision necessary to put it into the best form possible.

Aiken always wanted his students to go forward into new areas. As soon as a student received his degree and an instructor's position, he would be pushed into a course that did not necessarily reflect exactly his special competence and training. A graduate student or a young instructor might very well be called upon to teach a course he had never taken himself—perhaps because it was a new course, a course that didn't exist when he had been a student. Robert Ashenhurst remembers that he was asked to teach a course in numerical analysis, which Tony Oettinger too had been assigned to teach with no prior experience in the subject.

Fred Brooks recalls the laboratory setting of the 1950s. He shared an office with Kenneth Iverson. Aiken, who was Brooks's thesis adviser, would drop in every day "to read and discuss what had been written since the previous visit." It was, said Brooks, "quite a lot like having another father." Brooks too recalls Aiken's insistence that his students learn to express themselves "with clarity, with logic, with sensibility, with thought, but ever with clarity."

Two very different aspects of Aiken's personality come out strongly in students' recollections: the taskmaster teaching them how to think clearly and the concerned parent preparing them for the business of life.

By the mid 1950s Aiken was no longer known as "the Commander" (although he was obviously in command); he was referred to—behind his back, never to his face—as "the Boss" (and sometimes as "the Old Man"). When Fred Brooks arrived at the Comp Lab, Aiken was 53 years old, "at the height of his powers, alert, energetic, forceful, self-assured . . . and formidable," with a "Mephistophelian look." Maurice Wilkes relates in his autobiography how he found Aiken intimidating. Robert Campbell and Richard Bloch recall how Aiken would forcibly express his disdain for certain of his associates in the early days of Mark I, thinking nothing of humiliating one of his junior officers by assigning him duties below his rank.

On the other hand, Aiken treated anyone who had ideas, even ideas opposite to his own, with great kindness and respect. He respected spunk and did not suffer "yes men" gladly. Wilkes remembers the "look [that] would come in Aiken's eye," a veritable "look of invitation, almost of pleading, to an adversary, to close in for a good sparring match." An Wang, who became a member of Aiken's staff in 1948, reports that if Aiken "thought you were his peer, he would rein his temper and even tolerate being contradicted."

A regular feature of the Comp Lab was the daily coffee hour. Fred Brooks recalls how, every day, "at 5 o'clock in the machine room the crowd would gather" for "coffee and wide-ranging discussion." The members of the staff knew that this was where "business would be transacted." During one coffee hour, Brooks remembers, Aiken told Ken Iverson: "Next year you'll teach a course in business data processing.'" There had "never been a course in business data processing anywhere in the world," but "Ken was qualified because he was a mathematician." It was during a coffee hour that Aiken suggested to Brooks that "a payroll machine was the right topic for my thesis." Furthermore, it was during coffee hours that he discussed the history of the laboratory and the field, the strategies and directions of the field, and his philosophies about people, machines, and organization.

Aiken wanted every member of his team to get credit for his or her contribution, and he stressed the idea of team effort. Thus, he liked to have publications of the Comp Lab credited to "the staff of the Computation Laboratory" rather than to a single author. However, his name appeared separate from the rest in the list of the staff (which included students, associates, and even secretaries), and the preface carried his typeset signature. Sometimes, as in the 1949 Manual of Operations for Mark II, titled Description of a Relay Calculator, the staff was divided into "assistants" and full staff members. This practice went back to the Manual of Operations for Mark I, where

Comdr. Howard H. Aiken, USNR
*Officer in Charge*

was followed—in order of their naval rank—by the officers, then the enlisted men. The civilians were last.

At the 1983 National Computer Conference there was a heated discussion of whether Aiken's close control of everything published by the staff of the Comp Lab amounted to a form of censorship. Tony Oettinger insisted that one had to make a distinction between official publications of the laboratory, appearing with Aiken's name as director, and something that was an individual contribution that was "not part of the team effort." Aiken, according to Oettinger, "never censored anything" that was privately written by an individual.

Distinguishing between influence or suggestion and censorship, Fred Brooks found it "rather surprising" (in retrospect "a measure of the stature of the man") that, "considering his forcefulness, his dominant personality and his confidence in his own ideas," Aiken "nevertheless gave us an amazing amount of freedom, respected our results, our conclusions, and ultimately our writing." Brooks recalls having used a "particularly colorful analogy," involving dirty socks, in a draft of his thesis. After commenting "Well I wouldn't put dirty socks into *my* dissertation," Aiken said "It's *your* thesis, go right ahead."

Not everyone understood the distinction between private and official publications. Ronald King, of the Cruft electronics group, recalls vividly what happened when one member of Aiken's staff wanted to publish a book under his own name. According to King, the author "got into all sorts of hot water with Howard." King recalls Aiken saying: "Any member of the staff who publishes anything has got to have the whole staff on it, and my name comes at the top." The author was apparently just as stubborn as Aiken, which meant that they had come to a firm parting of the ways. Others, according to King, had "similar experiences—you had to toe the mark and be on track or you got fired." In contrast, An Wang called Aiken "an idealist," citing his belief that computer research "belonged in the public domain, and not to individuals or corporations." Wang, while working as a member of the Comp Lab staff on problems of memory related to the design of Mark IV, conceived the idea of using a doughnut-shaped magnetic core as a memory device. When he told his fellow staff members about his decision to patent this invention, they were "apprehensive" about Aiken's reaction, being aware of Aiken's passionate conviction "that computer developments should be kept in the public domain." After he filed his patent application, Wang was "a little nervous" about how Aiken would react. In the event, to Wang's surprise, Aiken "did not react at all," although he was certainly not "overjoyed." Shortly thereafter, Wang received a raise in salary, which he took as a sign that Aiken was "not too put out."

Although Aiken liked to have publications appear with the whole staff listed as a collective author, with his name standing apart and at the head of the list as "commander" or "director," Tony Oettinger has called Aiken "very modest about sharing in the credit, with the result that many publications, which in another environment might have had the professor's name leading all the rest, had his only in the acknowledgments." According to Oettinger, Aiken appeared to his students to be more interested in having ideas get "out in the open" so as to be "used in further investigation" than in personal credit. Oettinger has stressed that Aiken never "stole credit from a graduate student, which is . . . more than one can say for . . . a lot of faculty people."

The Comp Lab in Aiken's days had some of the features of a Renaissance atelier where a master artist (such as Verocchio) was in charge and where apprentices and senior craftsmen interacted with one another, a center where students learned their craft not only from formal instruction by a master but also from interchanges with one another and with artisans and visitors. In the Comp Lab there were always in residence one or more senior visitors—from Europe as well as the United States—spending anywhere from a few months to several years learning about computers and their possible uses. Intellectual exchanges with visitors from other countries and from other institutions were a part of the atmosphere of the Comp Lab. Students remember how the interchanges in the Comp Lab were an important part of their education and training.

Among the Europeans who came to consult with Aiken or to work with him and his staff for a brief or an extended period were C. Manneback (Belgium), José G. Santemases (Spain), A. Svoboda (Czechoslovakia), W. L. Van der Poel (Holland), A. Van Wijngeerden (Holland), Alwyn Walther (Germany), and Heinz Zemanek (Austria).

During the 1973 interview, Aiken told Tropp and me an amusing story about Van der Poel, who was a skilled weaver as well as a fine mathematician. During World War II, in Nazi-occupied Holland, Van der Poel used to weave small tea cloths with intricate designs, which he then sold to the German officers stationed in Holland. Through his contacts with the officers and his observations of the harbor, Van der Poel gained information concerning the movements of the German fleet. He encoded this information into the patterns of some special tea cloths which he was able to transmit to British naval intelligence. At the war's end, because of his fraternizing with German officers, Van der Poel was reviled as a traitor and collaborator, but this changed after the queen honored him for his wartime activity and made known what heroic work he had done. Aiken proudly displayed one of Van der Poel's tea cloths, a grim souvenir of the war.

Throughout the 1940s and the early 1950s, Aiken and his Comp Lab were at or near the forefront of computer science, but this position was eroded as the cutting edge of the new science and technology shifted to other academic centers and to industry. Despite the prestigious position of Aiken and the Comp Lab, it was becoming clear that new centers of excellence and innovation—notably MIT, Stanford, and Carnegie-Mellon—were gaining ground in this new field.

Aiken's renown as a pioneer continued to bring him medals, honorary degrees, and government decorations. His decorations and honors included the US Navy's Distinguished Public Service Award, the Air Force's Decoration for Exceptional Civilian Service, a Testimonial for Exceptional Civilian Services from the University of Wisconsin, an Officer's Cross of the Order of the Crown of Belgium, Palmes de l'Académie Française and designation as a Chevalier of the French Légion d'Honneur; in addition, he was elected a fellow of the Swedish Society of Letters and Sciences and a member of the Swedish Ingeniorsvetenskapsakademien, a Mitglied der Gesellschaft für Angewandte Mathematik und Mechanik in Germany, and a Consejo de Honor of the Spanish Consejo Superior de Investigaciones Científicas. In his curriculum vitae Aiken proudly listed among these international honors and his honorary degrees that he was a Registered Professional Engineer (license 7104) in the Commonwealth of Massachusetts.

Throughout most of the 1950s, Aiken was in demand as a speaker in America and in Europe. The record of his lecturing itinerary is dizzying. For example, between 16 August and 8 December 1955 he lectured in California, Wisconsin, Georgia, Sweden, and Germany. He spoke before large and small groups at engineering and business schools, before such organizations as the National Association of Machine Accountants and the gas and electric associations, and before such businesses as the Metropolitan Life Insurance Company and the Monroe Calculating Machine Company. There were also conferences at which he either was the main speaker or participated in the final summing up. This activity kept him in the limelight and ensured his prominent place in the computer scene.

Aiken generally did not write up his talks in advance. Indeed, he often (if not usually) did not write them up at all. He often spoke from a very detailed outline, often running to some 20 pages or so, in which each line was in fact a partial sentence. Without the straitjacket of a text he spoke somewhat freely, yet his presentation progressed in an orderly fashion from start to finish. He usually illustrated a lecture with a series of well-chosen slides. The result was an attractive combination of wit, erudition, and insight. Students remember him as a clear, forceful, and meticulous lecturer. He was always fully prepared and his lectures proceeded without hesitations or interruptions for the full period. During the 1973 interview I asked Aiken whether he had a file of his course lecture notes. I believed that these would provide a good index of his range of interests, his expectations of the students, and his changing view of the state of the science and the art. He assured me that he had kept nothing of this sort. Indeed, he went on to say, he always destroyed his lecture notes at the end of each academic year so that he would never be tempted merely to repeat the course of an earlier year.

During the 1950s, Aiken's reputation was maintained by the careers of his former students. In an obituary notice in May 1973, Robert Ashenhurst noted that Aiken's pupils included a former president of the Association for Computing Machinery and a former editor-in-chief of the *Communications of the ACM*, "two of the main architects of the IBM System/360, the inventor of one of the best known interactive programming languages, and others, including present chairmen of computer science departments." When Aiken was chosen for the Harry Goode award, the citation read, in part, "for the knowledge and inspiration imparted to many as a teacher."

The chairman of the Harry M. Goode Memorial Award Committee was Aiken's former student Isaac (Ike) N. Auerbach. Some remarks by Auerbach enable us to gauge the magnitude of Aiken's international reputation. Auerbach was one of the leading spirits in the organization of the first international computer conference (sponsored by UNESCO), the forerunner of the International Federation for Information Processing. According to Auerbach, most Europeans in the domain of computing believed it fitting that Aiken be the honorary president and, in that capacity, give both the opening and the closing address at the meeting. However, as Auerbach recorded, Aiken was not at first a supporter of the proposal. Yet, as Auerbach observed, "Europeans looked to Aiken as the biggest name in the United States that could come to Europe." Accordingly, Auerbach and Alton Householder (who was working closely with Auerbach on organizing the conference) convinced Aiken to take on the assignment. In his report on these events, Auerbach stressed the importance of Aiken in the minds of the European computer scientists. "If you were in Europe," Auerbach said, "the big name in the United States in the late 1950s was Aiken."

Despite Aiken's great reputation as a pioneer, a true founder of the new art and science of computing, it is a fact that in the late 1950s, because of his reactionary positions on many crucial issues of computer architecture and practice, he was no longer in any sense a leader. Although his students were exceptionally well trained in aspects of applied mathematics and other fields related to the computer, by the late 1950s there was a general feeling that Harvard's program suffered from being provincial. This feeling was shared by both insiders such as Fred Brooks and outside observers such as Maurice Wilkes.

Aiken was a maverick. He did not have the patience to develop a new area by persuasion and gradual development, by a delicate process of convincing colleagues and deans of the importance of a new area. At Harvard there are several different traditional paths by which new academic programs are inaugurated, new facilities are established, or existing programs or utilities may be enlarged. In the usual procedure, the first stage is to have a committee established by the Faculty of Arts and Sciences. This committee then produces a mission statement and a program. If this proves acceptable, the next step is to seek outside financial support and to apply for funds from the university budget. Sometimes, a special benefaction may produce an endowment, whose income can support a new endeavor. On occasion, the university's administrators may make certain special or unrestricted funds available for inaugurating some venture. The Kennelly bequest had made it possible for Harvard to establish a new Professorship of Applied Mathematics, Aiken's chair.

All these ways of starting computer science at Harvard would have involved some lessening of Aiken's personal authority as director. Furthermore, he must have known that he would not come out well in a general competition for university funds, since the administration and most of the faculty did not share his vision of the ever-increasing importance of the new art and science of computing. Aiken rather liked the Navy system, in which he was the Commander.

Aiken was aware of the cost of remaining "in charge"—of being able to make decisions without having to go through committees, deans, and votes of the faculty (save when absolutely necessary on questions of fundamental or major policy). The cost of independence was to have to seek support from outside the university to finance his own operation, to support his endeavors by personal persuasion outside the normal university circles. In 1949, Aiken estimated the cost to be $100,000 a year.

When the Navy's support of Aiken's activities drew to a close (after Mark II and Mark III had been constructed and delivered), he was able to gain some funding from the Air Force. As has been mentioned, the Air Force financed the construction of Mark IV and gave Aiken a contract to educate officers in the use of computers. Aiken also negotiated a contract with the Atomic Energy Commission. These funds enabled him to purchase the UNIVAC and thereby to bring his equipment up to the current state of technology.

The AEC contract ensured that there would be computer time on Mark IV for basic science in the university. Accordingly, the Dean of the Faculty established a committee to decide on priorities for proposals to use computer time. The committee's stated purpose was to judge and to order proposals in terms of their "intrinsic scientific value." Garrett Birkhoff has said that another reason why the dean had set up this committee was to try to monitor Aiken's activities. So great was the demand for computer time that in the autumn of 1949 the committee voted to accept no more proposals.

In 1949, Aiken drew up a historical memorandum describing the Comp Lab's achievements and sources of support and setting forth its needs for the future. A top priority was the construction or purchase of a new large computer, for exclusive university use, that would replace Mark I, which was now obsolete. He shrewdly noted that federal agencies were beginning to acquire their own computer facilities and could not be counted on for indefinite future support of Harvard's computer activity. In hard cash terms, the Comp Lab needed $40,000 to build or buy a new computer and also had to find a source for the difference between the annual running cost of $100,000 and the paltry $3000 supplied each year by the Department of Engineering Sciences and Applied Physics.

In the endeavor to seek capital funds to support the Comp Lab's needs, Harvard engaged the service of the Price Jones Company, a firm specializing in soliciting funds. The slogan writers at Price Jones came up with a plan and a slogan: "mass-production mathematics." The university rejected the plan because of "certain unfortunate dangers with respect to the university's internal uncertainty whether the Computation Laboratory is conducting scientific or industrial research." The essential ambiguity in Aiken's position at Harvard is evident here: Was his enterprise a proper one for a university like Harvard? Whatever the reason, the Price Jones affair came to nought and was abandoned.

Although as late as 1979 only about one fourth of Harvard's annual income came from federal grants and contracts (at a time when at Stanford and MIT the federal government provided slightly more than half the annual income), some $2.4 million of the total annual budget of $3 million for computer activity at Harvard came from federal funds.

There was considerable opposition to Aiken's academic "empire building," in part on the grounds that a computing center did not merit full academic status in a humanistic university but was rather like a service bureau. This ambiguity continued long after Aiken's departure from Harvard. In 1967, six years after Aiken's retirement, when it was suggested to T. Vincent Learson (president of IBM and a Harvard alumnus) that $10 million be raised for the Harvard computer program in the current capital fund drive, President Nathan M. Pusey cut the amount to a mere $1 million on the ground of "the desire of the central administration to keep a balance here among the objectives of the Program for Science."

At Harvard in those years, a somewhat similar and seemingly neverending debate focused on whether instruction on a musical instrument should be deemed a proper academic subject, with academic credit. Some members of the faculty considered such instruction as an applied or practical subject, one more fit for a practical music school than for a humanistic institution of higher learning.

We can see an example of Harvard's low priority for computer science and computing in an exchange of correspondence in 1951 between Aiken and Paul H. Buck, Provost and Dean of the Faculty of the Arts and Sciences. Aiken asked Buck whether Harvard would pay one-third of the operating costs of Mark IV in order to obtain one-third of its operating time for university services. Buck's reply was simple, straightforward, and negative.

Aiken was keenly aware of the tension that existed between him and the ruling forces of the university. In a humorous letter he wrote in 1950 to Harlow Shapley, pretending that he was Mark I ("Bessie the Bessel Engine"), Aiken decried "the absence of any more modern computing devices at Harvard University at the present time." The reason, he wrote, that "Bessie" was "still carrying the major burden of the work" was "in part the penuriousness of the institution." Aiken did get more computing power available when he was able to purchase a UNIVAC and when Mark IV, built for the Air Force, became operable and remained at Harvard, supported by Air Force funds.

These events draw our attention to yet another of Aiken's pioneering activities: finding peacetime support for science from the US government, particularly the Department of Defense and the Atomic Energy Commission. Today, government financial support for science and engineering is commonplace, and the annual operating budgets of universities depend in large measure on research grants and contracts from such agencies as the Department of Defense, the National Science Foundation, the National Endowment for Health, and the Atomic Energy Commission. In the post-World War II years, however, Aiken was a bold pioneer in negotiating financial support from the government and from the business community.

Even though Aiken was successful in finding the funds necessary to keep the Comp Lab going, he was aware that there was no easy solution to the problem of determining the proper place of the Comp Lab and the allied computer science program within the university structure. He could hardly help knowing that there was even some question of the propriety of an academic unit of the university operating a service entity such as a computer.

Aiken once "admitted" to Bob Campbell that he was not "in the 'classic' Harvard tradition" in either his "promotional activities" or his "close involvement with industry." He did have the support of certain "key people" at Harvard, he was glad to say, among them E. L. Chaffee. The existence of Mark I had resulted from Aiken's having been able to obtain major participation and a large expenditure from IBM, and he had phenomenal luck in showing how his work could promote "the interests of the Navy in better means of computation." But he knew that many at Harvard either were jealous of his success or questioned the propriety of having his business-oriented enterprise in a humanistic university.

Harvard's Committee on Applied Mathematics and the Computation Laboratory allegedly was created to resolve certain administrative problems and educational policies relating to the computer. It soon became apparent to some of the members that a major purpose of the committee was going to be to oversee Aiken's various activities and to try to bring them into better harmony with Harvard's established policies. Aiken, however, seems to have quickly undermined any such intentions of the committee. He wanted to be the sole arbiter of Comp

Lab policies. No one else was going to determine how his Comp Lab would be run or for what purposes the machines would be used. Aiken "just didn't know the meaning of cooperation," Garrett Birkhoff, a member of the committee, later recalled.

Aiken's difficulties with the administrative officers of Harvard arose from many sources, but a primary factor was the failure of the president and the dean of the faculty to recognize the growing importance of computer science as a subject worthy of a place in the university. Eager to get ahead with the business of training and teaching and of providing computer facilities to the university at large, Aiken was often frustrated beyond his powers of endurance by what must have seemed to him a lack of recognition and support from Harvard's administrative officers.[1]

1. The students who completed the Ph.D. under Aiken's direction were Herbert Francis Mitchell Jr. (1948), Miles Van Valzah Hayes (1950), Gerrit Anne Blaauw (1952), Charles Allerton Coolidge Jr. (1953), Robert Charles Minnick (1953), Kenneth Eugene Iverson (1954), Anthony Gervin Oettinger (1954), Lloyd Semon (1954), Theodore Singer (1954), Peter Calingaert (1955), Robert Lovett Ashenhurst (1956), Frederick Phillips Brooks Jr. (1956), Albert Lafayette Hopkins Jr. (1956), Roderick Gould (1957), Leroy Brown Martin Jr. (1958), and Gerard A. Salton (1958). The first two doctorates (Mitchell and Hayes) were officially awarded in Engineering Sciences and Applied Physics; all the rest were in Applied Mathematics. For further details, see *Makin' Numbers*.

# 24
## Retirement from Harvard

In 1961, Aiken decided to take advantage of Harvard's provision for early retirement and to begin a new career. He could have continued in his professorship for another five years, and possibly even for a few years after that. Instead, he chose to take advantage of a university rule that permitted tenured members of the faculty to retire at the age of 60. I don't know whether some particular event or series of events precipitated this decision. Mary Aiken recalls that he had a disagreement with some member of the university administration and decided that the time had come to start a new life. When I questioned former students and associates (including Bob Campbell, Fred Brooks, Tony Oettinger, Bob Burns, and Betty Jennings), they all agreed that the particular reason why Aiken retired so early was a mystery.

Tony Oettinger has written that Aiken had always said that he was at least as smart as most businessmen and wanted to prove that he was right. Fred Brooks concurs in this opinion. In a telephone conversation with Cuthbert Hurd, who was a close associate of Aiken's, especially after his retirement from Harvard, I was given confirmation of this reason for Aiken's having retired early from Harvard. Hurd said that he had never discussed this matter with Aiken, but that on two or three occasions when Aiken was in California, where he was a regular consultant for the Lockheed Missile and Space Division, the two of them had "talked at great length about organizing a company." "If we had done it and if it had been successful," Hurd mused, "it would have been the first microcomputer computer company in the world." Hurd told me that "an Assistant Director of Engineering at Lockheed . . . was doing the design work," and that "Howard, along with that man and me" would form the new company. Aiken, Hurd continued, "wanted me to help raise the money." They "never followed through" with this plan. "I thought that maybe he wanted to be rich," Hurd concluded, "and was thinking about starting the company for that reason."

I have always supposed that there was yet another reason for Aiken's shift from academia to business. The world of computer science and technology in 1960 was very different from what it had been 15 or even 10 years earlier. It had become compartmentalized into highly specialized technological domains. In addition, the state of knowledge and the state of the art were changing at an extraordinarily rapid rate. It had to be obvious to anyone as intelligent as Aiken that he no longer was a "top dog" in the world of computing, that he wasn't any longer at the forefront of the subject he had worked so hard to create. He was not the kind of person who could live comfortably in the role of a "has-been." And so, I believe, it would have been quite in character for Aiken to quit while he was still "ahead."

In certain respects, Aiken had rapidly become a conservative figure in the world of computing. In the 1950s he was already "old" by the standards of this rapidly advancing science, art, and technology. Computer science and invention had become a young man's game. Even in the years just after the war, many of the major advances had come from young men trained in the new electronics of radar rather than in classical electrical engineering. In the words of Maurice Wilkes, the new computer innovators were young men with "green fingers for electronic circuits," many of whom had come from experience with radar and "were used to wide bandwidths and short pulses."

Although Aiken's field of science for his doctorate was electron physics, and although the subject of his dissertation was space charge within vacuum tubes (or electron tubes), his expertise was in the physics of vacuum tubes and not in electronics, not in the design and application of circuits using vacuum tubes. Indeed, Aiken had not taken the course in vacuum tubes that had first been given at the University of Wisconsin when he was in his senior year.

Wilkes, whose esteem for Aiken's early achievements has been cited on a number of occasions in this volume, has written that he could not help but have the feeling that Aiken, in his last years at Harvard, was no longer in a "position of leadership." Even Fred Brooks, one of Aiken's most devoted disciples, admitted in retrospect that the Harvard Comp Lab had a "Charles River view of the world"—that the students were not fully aware of the developments taking place in other institutions (including MIT, just down the river).

In discussing Aiken's early retirement, Gerard Salton, one of Aiken's last Ph.D. students and later a professor of computer science at Cornell

University, suggests another factor that may have led to Aiken's decision. After a decade and a half of struggle with Harvard's central administration (in the last years, under President Nathan Marsh Pusey and Dean of the Faculty McGeorge Bundy), and also with the administrative deans of the Division of Engineering and Applied Physics, Aiken may simply have become weary. But Salton takes note of another factor. For most of the years of the Comp Lab and the graduate program, a candidate had had to meet Aiken's personal approval in order to succeed. In the 1950s, however, Aiken became a victim of his own success. His students became junior faculty members, and they began to enjoy personal and professional contacts outside the immediate circle of the Comp Lab. The new faculty members began to have independent projects and even began to attract graduate students, whereas in the past all such students worked directly under Aiken. Students, according to Salton, chose to work with the new faculty members, who were "more easily approachable than Aiken," and, as a result, "new students no longer flocked to Aiken's door." In short, the old Comp Lab, centering around and controlled by Aiken, had ceased to exist. Aiken, surely aware of this change, realized that it was time to retire.

The evidence supports Salton's analysis. During Aiken's last years at the Comp Lab, for the first time, doctoral dissertations were completed under the direction of the junior staff rather than under Aiken's immediate supervision. Three Ph.D. degrees were awarded in 1960, and Aiken was not the thesis director for any of them. One of the trio, Derek Henderson, wrote his dissertation, "Logical Designs for Arithmetic Units," under Peter Calingaert; Aiken did not even serve as a member of Henderson's doctoral committee. Neither of the two others who received the Ph.D. degree in 1960—Ramón Alonso ("Digital Calculators with Solution of Ordinary Differential Equations") and Martin Cohen ("Switching Functions over Integer Fields")—wrote his dissertation directly under Aiken. A little more than 10 years later, Aiken, in retirement, nevertheless considered these three to have been his students. During the 1973 interview, he showed Hank Tropp the three dissertations, which were among the others that he had placed with pride in a stately row on the mantel shelf of his living room.

On 6 October 1961, some months after Aiken's retirement from Harvard, there was a surprise testimonial dinner at the Harvard Club of Boston. Described as a gathering of his friends, associates, colleagues, former students, and "leaders in the design and use of large computers in science, industry, and government," the dinner was arranged by Tony Oettinger. Aiken gave an informal talk, apparently speaking from memory; I have found no record of what he actually said. There is no doubt in the minds of those who were present that the high point of the occasion was the presentation to Aiken of a silver model of his favorite computer, Mark I. The formal presentation was made by Charles A. Coolidge, Acting President of Harvard University and father of one of Aiken's Ph.D. students, Charles A. Coolidge Jr. The delicately crafted model, about 2 feet long and 6 inches high, was commissioned by Tony Oettinger and was built by Reino Martin, a silversmith in Annisquam, Massachusetts. It was exact, even down to such details as tiny flashing panel lights and tiny electric typewriters with silver keys and silver ribbons.[1] Later in the academic year, at Oettinger's urging, Harvard honored Aiken in two ways which are the summit of possible recognition of distinguished faculty members. At the next commencement, Aiken was awarded an honorary doctorate. Then the Comp Lab was formally renamed the Howard Hathaway Aiken Laboratory of Computer Science.

1. During the 1973 interview, Aiken proudly showed us this model, reposing on a mantel shelf in his living room alongside the bound copies of the theses of his doctoral students.

# 25

## Businessman and Consultant

After retiring, Aiken moved to Fort Lauderdale, where he entered the world of business, founding and running his own investment company, Aiken Industries. Essentially, Aiken Industries specialized in taking over companies that were ailing and bringing them back to good health, at which point they were sold. He could not help but be active at the university level, and he accepted a part-time teaching post at the nearby University of Miami, becoming a colleague of John Curtiss.

I was able to learn more about Aiken's activities during the years after his retirement from Harvard from Richard McGrath, an attorney who was closely associated with Aiken from 1962 to 1973. During these years, McGrath—accompanied by Martin Flaherty and James Marsh—was "almost constantly on the road" with Aiken, "helping him to build his company (which was originally known as Howard Aiken Industries Inc. but was later called Norlin Technologies Inc.)." The three of them "assisted" Aiken in the "process of acquiring companies that became divisions of Aiken Industries." According to McGrath, Flaherty and Marsh "then managed several of the companies for Howard" while he served in a legal capacity.

McGrath remembers Aiken as "a born teacher and mentor" who "left an indelible imprint on all our lives." Like all "great teachers," McGrath adds, he "demanded that each of us perform to the limit of our abilities." He "seldom spared our feelings or those of others in the quest for truth," McGrath concludes, but McGrath's "principal memory" of Aiken is "as a kind, gentle, thoughtful, and unreservedly loyal person who demanded (and generally received) unreserved loyalty in return." What McGrath found particularly noteworthy was Aiken's "management style," the way in which "he put together a highly successful high technology company with multiple divisions, without ever having an office or secretary of his own." Aiken, McGrath concluded, "literally worked out of his hat," perfecting "the art of visitation, traveling from division to division like a visiting fireman."

During the 1973 interview, Hank Tropp questioned Aiken about aspects of his life and career after leaving Harvard. Aiken referred, first of all, to his "forming Aiken Industries, beginning in 1961" and his becoming "vice-chairman of the board" in 1967. "So now," Aiken said, "I go to board meetings, but I'm not going at it the way I used to. . . . When they kicked me out of Harvard, I had to find a new job and that was Aiken Industries. And when *they* kicked me out, I had to find a new job and went into the consulting business. So now I spend a good deal of time at Monsanto." In reply to a question from Hank, Aiken said that he had been a consultant at Lockheed "for many years," but that he had "quit that this year."

Once the subject had been brought up, Aiken felt the need to discuss the subject at length. "You can't quit," he said. "At the time you quit, you've had it." "If I were to quit work and sit here in this study," he continued, "I think I'd be dead very soon. I don't think I'd last."

Obviously, someone as deeply committed to computers as Aiken, endowed with a continually creative mind, could not completely forsake the field of activity in which he had been so prominent. During his last years, Aiken continued his long-time service for the aerospace industry in California, making periodic visits to Lockheed Missiles Company. Cuthbert Hurd gave me a list of some people with whom Aiken was associated in California when he used to come out to the West Coast on his regular visits. One of them was George Garrett, who, Hurd informed me, "was for a while the Director of Computer Activities at Lockheed Missiles. Howard and George saw a great deal of each other for a certain period." Another person at Lockheed with whom Aiken was closely associated was David Willis, "a key person" in organizing for Lockheed the "worldwide" symposium on switching in relation to the aerospace industry. Hurd remembered that Jack Nash and David Willis had worked with Aiken on that meeting. (Hurd had in mind the symposium on switching theory, the proceedings of which were edited by Aiken and William Main.) Hurd told me that at that time Willis "lived in Menlo Park and frequently, when Howard came [out] here, he would have dinner at the Willises because Mrs. Willis was not only beautiful, but a wonderful cook."

Hurd, on another occasion, gave some further details on a new company that Aiken had proposed to form. He recalled that he, Aiken, and William Main had met several times in 1970 "to discuss the formation of a corporation to be called PANDATA." The three had "discussed the idea of what was to become a microprocessor and personal computer." Aiken, according to Hurd, "had an early vision of the usefulness of such devices," "believed that they could be mass produced at a low cost," and "wished to form an integrated company to manufacture and sell them." Aiken wanted Hurd "to help form the company, be chairman of the board, and raise the money." Aiken himself "wished to make a considerable investment in the new company." Hurd reported, however, that he "was busy at the time with other activities" and that Aiken "died before the venture could be launched."

One of Aiken's final computer-related assignments in these post-retirement years was for the Monsanto Chemical Company, which was trying to develop magnetic bubbles as the basis of a new memory technology. At the time of the 1973 interview, Aiken was enthusiastic about magnetic-bubble technology. This led him to talk about miniaturization in general. He had on the table a hand-held Bowman electronic calculator, and he showed an obvious sense of delight as he discussed how powerful a tool this small device was. He said that he foresaw a time when a machine the size of this calculator would be more powerful than mainframe computers. Aiken had never been concerned to patent his innovations while a member of the Harvard faculty, and the innovations he produced for Monsanto were patented in the company's name rather than his. But he was concerned with a patent for one of his own inventions, the creation of his last retirement years. The invention in question was related to the general problems of encryption and decoding and the security of computer data.

Aiken went into some detail about this most recent invention and the company that was in the process of being organized to exploit his innovations in relation to the security of computer information. "That's the Information Security Corporation," Aiken said. It was "being formed by Dick Bloch to exploit a cryptographic invention of mine." The parent company was called Genesis. In an earlier interview with Bloch, Tropp discovered that the primary mission of Genesis was to seek new ideas that could be exploited commercially and then to find financing. Once the venture was started, Genesis would provide the early management; as soon as the company was able to stand on its own feet, however, Bloch and Genesis would, in a sense, "get out and look for something else." "Is that about right?" Tropp asked Aiken. Aiken replied, "Yes. Except that in this case, this is especially close to what Dick wants to do, so that I doubt that he'll leave it. I think he'll stay in this company for quite a while." Aiken observed that an "interesting thing in all of these discussions" was that "you see the same names keep cropping up." Dick Bloch, for example, "was one of the first officers on my staff when the Mark I started operation. Now, that's all over and done but here's another association." He was rather proud that "the men that I was associated with in the beginning at Harvard have not just run away like last winter's snow, at all. There has been a certain cohesiveness."

The new company being formed to exploit Aiken's ideas on encryption encountered a severe block when an attempt was made to patent Aiken's innovation so that it could be put into commercial practice. Apparently, according to Cuthbert Hurd's recollections, "on the day the patent was issued, the US Department of Defense classified it as 'Secret' and confiscated it." This innovation, based on Aiken's knowledge of number theory, was considered to have implications that were too important to allow it to be given general circulation.

Aiken's solution of the problem of encryption was ingenious and displayed his background as an applied mathematician. Basically, as Dick Bloch explained the principle of Aiken's invention to me, his proposal made use of a clever way of generating random sequences. A coded key would enable a user to generate the same sequence that had been used in the stage of encryption and thus read whatever message or information had been encrypted. Even if someone were to find the key, it would not be possible to "break" the code unless that person also knew the particular way number theory was being used to generate the proper random sequence. Aiken finally got his patent, but his invention was never fully exploited, chiefly because of his early death. Yet it still brings in royalties to his estate.

When Aiken found that he was being barred from fully exploiting his invention, he decided to bring a suit against the government for the loss of sales. For this purpose, he needed to know the number of computers in the United States, on the basis of which he could compute the number of potential users of his mode of encryption. His subsequent actions had one very amusing consequence, which led to the occasion of the 1973 interview.

The interview was part of a program inaugurated by AFIPS and the Smithsonian Institution, as has been mentioned earlier. The goal was to record the life histories of computer pioneers. Hank Tropp, with whom I had gone to Fort Lauderdale to interview Aiken, was the program director. A professor of mathematics at Humboldt University, Tropp had led a team which produced an amazing collection of primary source material that documents the dawn of the computer age. Future historians will be grateful to AFIPS and the Smithsonian for having sponsored this notable historical endeavor and they will be aware of the skill and insight with which Hank Tropp elicited from these computer pioneers their recollections of their activities and even obtained crucial documents which are kept together with the tapes and transcripts. My association with this project was minimal. I had been asked by Walter Carlson, then president of ACM, to serve as a kind of informal adviser. The reason was that I was then one of the very few professional historians of science interested in computer history. Tropp especially wanted my help with regard to an Aiken interview because he tried in vain to get Aiken to agree to a recording session. Tropp hoped that I could be more persuasive than he himself and others had been, that Aiken would agree to an interview if I were to conduct it in association with Tropp. After all, he reasoned, I did know Aiken, I had been a colleague, and I was a historian.

Accordingly I wrote several letters to Aiken, urging him to allow Tropp and me to interview him. Each time he would put me off. Finally, in desperation, I wrote him a long letter, calling on all the eloquence I could summon, telling him how especially important it was for him to give us an interview. I stressed the fact that his significant role in the birth of the computer was not being recognized, primarily because Mark I had not been electronic. I urged him to see us and to set forth his own side of the story, to help ensure that his place in history would be carefully defined and preserved. Aiken had been ill and had recently undergone surgery at Johns Hopkins. Soon after receiving my letter, he did at last agree to the interview. In my innocence and vanity, I assumed that the seriousness of the operation and the eloquence of my appeal had won him over. It was only some years later that I learned that the reason why Aiken finally gave in and permitted us to have our interview was not simply the result of my persuasive powers and my eloquence. The occasion for my learning why Aiken finally allowed us to do our interview was an interview which I was conducting with Cuthbert Hurd at his home near Palo Alto.

Hurd told me that when Aiken learned that his patent on encryption and the security of data was being blocked, he called Hurd to get some information of importance in building up a claim against the government for financial loss. He wanted to know, Hurd recalled, how many computers there were in operation in the United States. "With your contacts with IBM," Aiken said, "you should be able to find that out for me." He wanted Hurd to "please ask IBM how many tape and drum and disk units it expected to ship in the next few years."

Now it happened that Hurd was a friend of Hank Tropp and of Walt Carlson, with whom he had been associated ever since their days together at IBM. Furthermore, Hurd had a long sustained interest in the history of the computer. He was a strong supporter of the oral-history project and he was concerned that we had not been able to pin Aiken down for an interview. Accordingly, he seized this opportunity to get Aiken to agree to be interviewed. He told Aiken that he would find out "how to obtain the information he wanted" (which, he told me with a smile, actually could "be obtained from public sources") if Aiken would agree to be interviewed by Hank Tropp and I. B. Cohen. Aiken accepted this "quid pro quo" and agreed to allow Hank Tropp and me to conduct an oral-history interview with him. Accordingly Hank Tropp and I went to Fort Lauderdale for two memorable days of recorded discussions in February 1973. From start to finish, Aiken was affable, considerate, and friendly—to a degree that belied his reputation for fierceness.

# 26
# *A Summing Up*

Howard Aiken died in his sleep in a hotel in St. Louis on 14 March 1973. He was in St. Louis for one of his regular consultations with Monsanto. He was 73 years old.

Despite the many frustrations of his Harvard years, Aiken was always appreciative of the important role that Harvard had played in his career, not only in providing his graduate education and teaching opportunity for some three decades but also in supporting his original proposal for a super-calculator and providing an academic home for the machine once it had been constructed. He was proud that Harvard had bestowed on him an honorary degree and had named the computation laboratory where he had spent many creative years after him. The terms of his will provided for the establishment of a trust, administered by Richard McGrath (the lawyer who had worked with him in establishing Aiken Industries), of which Harvard University is the residual beneficiary.

Although Aiken was widely known for his acerbic style, he could also be a kind and thoughtful human being, truly concerned about the welfare of others. This is the reason why so many former students, colleagues, employees, and associates remained loyal friends and cherish warm memories of their relations with him while nevertheless being aware of his harsher side and even believing that he treated some people too severely or unfairly.

The warm, humane side of Aiken's personality came to the fore in many ways. Kenneth Iverson remembers Aiken as spending "very little time in his office and much time rolling around the lab, speaking to people at every level about their work." He always had a friendly word to say to "draftsmen, machinists, typists, and janitors."

Jacqueline Sill, Aiken's long-time secretary, recalled the contrasting aspects of Aiken's character. Although he was a "hard driver," she reported, he always had in mind "the benefit of the driven." By contrast, she noted, he "quenched the presumptuous with cruel workloads" and he "used mock and real explosions to bring out the most any person could give." He would use "derisive expletives" in order to make "students rethink their work." He was fond of saying "I am a simple man and I want simple answers," and so he guided his staff to straightforward responses and non-convoluted reports. This quality of Aiken as a teacher has also been stressed by Tony Oettinger, who said that Aiken "brought more out of people than they knew they had in them."

Jacqueline Sill insisted that Aiken could be a "patient, considerate, and compassionate guide" to his students in a way that contradicts the often forbidding aspect of his appearance, language, and behavior. She used three adjectives to describe him: "human," "humane," and "patriotic." He "was extraordinarily generous," she said, noting that "few people" were aware that in almost "every instance in which he received an honorarium for giving a speech, that money was put into what we at the Lab termed the Comp Lab Welfare Fund." This "money went to subsidize gifts to staff members when they married" or when they "had children," provided "condolences to staff members at the times of need for such things, loans to help someone tide over a period between paychecks, and other such charitable purposes." It was characteristic of Aiken and the persona that he cultivated that "he never wanted the source known." After his death, Jacqueline Sill believed it "appropriate to break that confidence." She pointed out that "outsiders knew only the stern, serious, dedicated man," but "insiders were aware of a quite different Aiken, one who "was funny," who "loved and appreciated jokes, whether he was the teller or the butt," who played tricks and who was the subject of many practical jokes. He could readily be either "the comedian *or* the 'straight man.'" On the occasion of the memorial service in the Harvard chapel, she recalled "this aspect of Aiken, his great humanity and his concern for the welfare and careers of his students and associates" and remarked that it was such aspects of his personality that caused many of his students, colleagues, friends, and associates "to remember him with such deep affection."

An anecdote related by Richard Bloch shows Aiken's deep concern for his associates. There was a time during the war when pressure to get out results was so great that Bloch had to work all through the night in an attempt to finish the job on time. When Aiken arrived in the morning, he recognized at once that Bloch was so fatigued that he could scarcely continue to function normally. Bloch wanted to go on working, but Aiken knew that Bloch had reached the limit of endurance and needed to go to bed. So, as his commanding officer, he ordered Bloch to go to his sleeping quarters, which were in the Hotel Commander. Filled with concern, Aiken wanted to be sure that Bloch actually did go to bed to get the sleep he so desperately needed. Accordingly, he personally escorted Bloch across the Law School Quadrangle and the Cambridge Common from the physics laboratory where Mark I was housed to the hotel. He even waited outside the room until Bloch got into his pajamas and his bed. Only then did Aiken bid him have a good sleep and then return to his computer and the day's work.

I have mentioned earlier Fred Brooks's remark about Aiken's dual character—that although Aiken was "an admirable man," he was "not unwilling to unleash his full powers on people who were 'smaller' than he was." In his recollections of Aiken, Brooks also called attention to a quality noted by Maurice Wilkes. Aiken "liked spunk," Brooks recalled. "Those of us who were refractory and difficult got along well with him." Brooks remembered that he and other graduate students considered Aiken as if he were "another father."

When Aiken was honored posthumously at the National Computer Conference in 1983, many former students, associates, employees, and colleagues spoke warmly of his human qualities while also recognizing the harsh side of his personality. In preparing this portrait, I was pleasantly surprised to find how many of those who knew him still cherish warm recollections of his generosity and his overall kindness as well as his stern disposition.

As Jacqueline Sill recalled, Aiken had a great sense of humor. He enjoyed both playing jokes on others and having jokes played on him. One day when Aiken arrived at the Comp Lab he was furious to discover that the computer had not been running all the previous night. There was a chart on the wall which indicated either by a blue line that the machine had been running regularly all through the night or by a red line that it had stopped. He was obsessive about having Mark I run smoothly and constantly day and night and so, on his arrival each day, the first thing he did was to look at the chart. If the line was blue, he would say good morning to the staff; but if the line was red, he would storm off to find out what had gone wrong and who was responsible.

Bob Burns, who worked with Aiken for many years, relates how one morning Aiken arrived to find a solid red line. "What the hell's going on here!" he shouted at the operator who was on duty. "What have you been doing? Who have you called? Where the hell is Hawkins?" Hawkins was Bob Hawkins, the inveterate smoker of a corn-cob pipe who had learned at IBM's plant in Endicott how to keep Mark I running.

When the operator told Aiken that Hawkins was downstairs, Aiken rushed down to find him. There he was in his office, calmly reading the morning paper and drinking a cup of coffee as if he had no care in the world. Aiken could hardly contain himself. "What the hell are you doing down here reading the paper?" he shouted at the top of his voice, "Why aren't you upstairs? The goddam machine's been broken down for thirteen hours!"

"You're crazy," Hawkins retorted in an even voice. "The machine ran all night long." Aiken replied that the chart was red and he hurried back upstairs with Hawkins close behind him. When they got to the chart, they saw an operator slowly peeling off the strip of red tape that had covered the blue line on the chart. Everyone who was around burst into laughter. "Well, I guess I've been had," said Aiken, grinning. The staff fashioned a large red badge in honor of the occasion and Aiken wore it proudly all day long. Jackie Sills long remembered Aiken's "twinkling eyes when he perpetrated a trick, his 'straight man' look when he was the butt of a joke, his uproarious laugh when something struck him funny."

Engineer, inventor, applied mathematician, and entrepreneur, Aiken stood four-square for honesty, openness, and plain dealing. He tended to be self-willed, stubborn, and difficult but never devious. He meant every word of his response to Thomas Watson, junior, when he refused to sign the non-disclosure agreement. It was his moral code. He did not believe that university professors should take out and hold patents; he wanted all knowledge to be open and free. Aiken's name did appear on the IBM patent for Mark I, but until late in life, when he became a businessman and professional inventor, he had no interest in patents or—for that matter—priority squabbles. Fred Brooks recalls a discussion of patents that occurred during one of the coffee hour sessions at the Comp Lab. Aiken was asked point-blank why he had not challenged the patent on the magnetic drum taken out by Engineering Research Associates, even though he was "certain of his own priority in the invention." His answer was that in the computer field the problem "was not to keep people from stealing your ideas, but to make them steal them." Brooks notes that Aiken "was never consumed with a passion for credit."

Aiken was always proud of having been an officer in the Navy. Grace Hopper recalls Aiken's bitter disappointment that the Navy never promoted him above the rank of Commander. She also expressed concern, when she was promoted to the rank of Captain, that she now outranked her former Commander.

Aiken was always conscious of national security problems, recognizing the difference between military and academic affairs. Colonel George R. Weinbrenner, one of Aiken's many friends in the military, recalled at the Aiken memorial service how Aiken "cautioned against complacency in research and development for national defense." Aiken, he reminded us, had been "instrumental in bringing to our military R&D programs a scientific discipline that has permitted this nation to maintain itself as a world power."

Aiken was hard on himself and on others because of the high standards he set for himself and for anyone with whom he was in contact. And yet he could also be generous and forgiving. Toward the end of our interview in 1973, Aiken suddenly became aware that he had spoken freely about himself and others without any reservation. He recognized that every comment of his had been recorded on tape—permanently, for the public record, for history. During the interview he had made some especially disparaging remarks about two colleagues, one at Harvard and the other at MIT, and he decided that he did not want these slurs to remain as part of the permanent record. And so, when we reached the end of the day's taping session, he indicated that he had a further remark to be recorded. Then, in a clear and unambiguous voice, he directed us to erase his unkind remarks about two individuals. He did not want his slurs on these men's reputations to go down in history as part of the record.

Aiken was married three times. The first marriage was very short lived, ending in an annulment. On 7 January 1943 he married Agnes Montgomery, known to their friends as Monty. She was a high-school teacher of French. They had a daughter, Elizabeth. Betty Jennings and others remember occasions when Elizabeth Aiken would come to the Comp Lab. Shortly before Aiken retired from Harvard, the marriage ended in divorce. In 1963, he married Mary McFarland, who had been a teacher in the elementary schools of Boston. Although I had come to know Aiken as a junior colleague, I did not know the Aikens socially, and I never had the occasion to meet Monty. I did get to know Mary during the course of the time that Hank Tropp and I spent in Florida conducting the oral history interview; she and I have continued to be friends in the succeeding years.

Howard and Mary were both high spirited and strong personalities, but they obviously were having a good life as husband and wife. Both of them enjoyed intellectual sparring and there was never a dull moment in their lives together. Mary Aiken was not trained in mathematics and was not primarily concerned with the actual technology of the computer, but she was—and remains—a fierce champion of Howard's greatness and his important place in history. After Aiken's death, Mary summed up his life as "colorful, inventive, stormy, and changing." In his final years, Aiken was very interested in the careers of his stepdaughters, who in turn were very fond of him.

After Aiken's death, the warmth of Aiken as a family man was summed up beautifully by Thomas Leonard, his old college roommate. "Myrtle and I always treasured his visits here," Leonard wrote to Mary at the close of a letter of condolence, "and especially when you came with him." He recalled to Mary how she had "contributed a most important part of his life" and expressed the hope that she would remain "part of ours." Our "hope and our friendship," he concluded, "will be for you the same as it had always been for Howard."

Aiken had wise insights into many aspects of technology and the influence of changing technology on our lives. During our meetings in Fort Lauderdale, we spoke of the computer as a labor-saving device. Aiken then explained to us that there was a very fundamental misconception about all labor-saving or time-saving innovations. It was commonly believed, he said, that the result of such new technologies would be to give human beings more time for leisure. In fact, however, the computer would not bring more leisure but would merely make human beings more productive during their work hours. The net result would be that we would all work harder during hours at the job.

In one of his many lectures, Aiken stressed this misconception of inventions as labor-saving devices. "It is my belief," he said, "that there has *never been a labor-saving invention made yet. Never!*" And then he added, "I don't think there will ever be." He explained what he meant. "The automobile," he said, "was going to be a labor-saving invention. You were going to save time in traveling." In practice, however, "what you did, you made the automobile so pleasant to travel in that you now travel ten times more than you ever did before." "And it's the same thing with this computer," he said. "We're going to save labor, we're going to change the distribution of labor, we're going to change the distribution of costs, in this way improve the standard of living for all." But, he commented, "we *don't* eliminate people, they aren't obsolete yet." In this final prediction, however, Aiken has proved to have been only partly right. Certainly, people who use computers in their work actually work more efficiently, but they also work harder and even for longer hours than formerly. But there is a total saving in labor in that the introduction of computers has eliminated many jobs that formerly were performed by human beings.

Aiken's last public address before retirement from Harvard was given at the International Congress on Automation, held at Madrid in 1961. The organizer was José Santamesas, a Spanish professor of electrical engineering who had spent some time at the Harvard Comp Lab. Aiken's presentation was the final act of the congress, of which he was the honorary president. "History has made it extremely clear," said Aiken, "that the effect of automatic machines has been and [will] continue to be that we produce a greater output per worker." That is, he explained, we "make use of a greater installed horsepower per worker." The "result" is that we "provide an improved standard of living for the worker and, above all, we eliminate drudgery and boredom."

All his life, Aiken was fond of making predictions. Grace Hopper once told me of a day when Aiken announced to the staff that the time would come when the equivalent of the whole computer power of Mark I would be fitted into a large shoebox. She and the other members of the staff did not know on this occasion, as on similar occasions, whether Aiken was being serious or was having them on. One of Aiken's most famous predictions, the one most cited in books on the computer, is that only a few computers would be required to serve the total needs of the country. Sometimes the number is given as six or a dozen and sometimes even two or three. The evidence shows, however, that this particular statement was made in the context of certain proposed large-scale computers being planned for the Bureau of the Census through the National Bureau of Standards. It was being proposed that the Bureau of Standards should plan to have five such machines built, but Aiken concluded that the potential work load would not require more than two.

During our visit with Aiken in Florida in 1973, just weeks before his death, he made several predictions. One was that in "another five years . . . every kid is going to have a Mark I." This prediction, like the one about the shoebox recorded by Grace Hopper, rapidly came true (though not quite so rapidly as Aiken forecast). Another seemed to me to be preposterous. We were talking about costs and I remarked that I was greatly impressed by the fact that as the size of computers was diminished and their computing power increased, the cost of computing went down. I observed that this was one of the few technologies—if not the only one within my ken—where technological capacity was improved at so dramatic a rate while the cost went correspondingly down and not up. Aiken agreed, but he reminded us that while the cost of hardware was going down, the cost of software was going up. And then Aiken gave us a concluding prediction, based on an extrapolation of the cost curves to their ultimate limit. "The time will come," he said, "when manufacturers will give away computers so as to be able to sell software." When I mentioned this prediction at a meeting a few years ago, the commentator on my paper remarked that he had just purchased a new computer; it had not been given away in order to sell software. Anyone who has recently bought a computer will agree with this commentator and will appreciate that Aiken's final prediction is, as of today, still far off the mark. Only the future will tell whether in this case, as in many others, Aiken had the gift to foresee the future correctly.

In an article in *Business Week* (6 March 1995), however, there is discussed an example in which Aiken's prediction has almost been realized. According to *Business Week,* Hiroshi Yamauchi, president of Nintendo Co., the "world's leading video game company," declared, "There's no way to charge a premium on hardware." Accordingly, Nintendo "charted a business model in which the game consoles would be 'given away' to consumers at cost, or even below, to boost sales of software." This is almost Aiken's prediction come true.

Few if any of the Harvard students and faculty of today have any sense of either the monumental quality of the technical achievement that Mark I represented or of its role in formally opening the computer age. Aiken's role as a pioneer in the emerging world of computer science is no longer part of the common knowledge of the computer community.

Aiken's important position in the new world of computers was admirably summarized in 1964, in the citation for the Harry Goode Memorial Award of the American Federation of Information Processing Societies. This citation expressed the judgment of Aiken's peers, of his fellow computer scientists who composed the award committee; the chairman was Aiken's former pupil, "Ike" Auerbach, and a good friend, Alton Householder, was a member. The first part of this citation read

—for his original contribution to the development of automatic computers that led to the first large-scale, general purpose, automatic digital computer ever to be put into operation.[1]

The second part read

—for his continuous work in the field of digital computers as an engineer.

It is far from clear what the committee intended by the phrase "as an engineer," but very likely they recognized that, despite the acknowledged seminal role of Mark I, Aiken's importance was greater than as a designer of machines or as an inventor of software. The common definition of an engineer is a person who applies scientific and mathematical principles to practical ends. Hence, the phrase "as an engineer" may have been intended to characterize Aiken's endeavors to find continuing new applications for the computer—notably in data processing rather than scientific calculation, in automatic billing, and in production control and other non-numerical uses. The committee were certainly aware of Aiken's important contribution in these domains.

The third and final part of the citation read

—and for the knowledge and inspiration imparted to many as a teacher.

In retrospect, his inauguration of a program for training in the emerging field of computer science may indeed have been one of his greatest pioneering achievements. The galaxy of students who received degrees under his direction may, in the long run, have been his greatest contribution to the developing computer age.

In a final assessment of Aiken's role in history, it would be hard to do better than the Harry Goode Memorial Award committee. Aiken's position in the history of the computer is difficult to assess because he is not celebrated today for a specific invention or innovation such as the stored program or the magnetic core or the chip. But his tripartite

1. Since that time we have learned more about the Colossus machines at Bletchley Park and the machines of Konrad Zuse, and today's historians would question the words "first . . . ever." But at that time, there would have been a rather general agreement with the sentiment being expressed: that Aiken's brainchild had played a seminal role in opening the computer age.

role in establishing the computer age, recognized by his contemporaries, was nonetheless a major factor in producing the computer world in which we live. This is, indeed, a monument "aeris perennius," far more lasting than brass and even stones and mortar.

At the banquet honoring him at the time of his retirement, Aiken expressed a sentiment that all readers of good will are likely to applaud. "I hope to God," he said, that the computer "will be used for the benefit of mankind and not for its detriment."

# *Appendixes*

# Appendix A
# The Harvard News Release

*This appendix reproduces the original news release issued by Harvard University concerning the new machine. After the official unveiling and presentation of the machine to Harvard, and after Thomas J. Watson had been angered by the original news stories, the News Office issued a revised version that gave more credit to the IBM engineers. Where the original release presented Aiken as "the inventor" who had "worked out the theory which made the machine possible," the revised version said that Aiken "originated the basic theory which led IBM to assign its engineers to invent a machine which would carry out the calculations desired by Aiken."*

HARVARD UNIVERSITY
Cambridge Mass.

university news office

RELEASE: Monday Papers
August 7, 1944

> (The NAVY, which has sole use of the machine, has approved this story and set this release date.)

World's greatest
mathematical calculator[1]

The world's greatest mathematical calculating machine, a revolutionary new electrical device of major importance to the war effort, will be presented today to Harvard University by the International Business Machines Corporation to be used by the Navy for the duration.

---

1. The revised version has an additional line: "I.B.M. automatic sequence controlled calculator."

This apparatus will explore vast fields in pure mathematics and in all sciences previously barred by excessively intricate and time-consuming calculations, for it will automatically, rapidly, and accurately produce the answer of innumerable problems which have defied solution.

The ceremonies in University Hall today will include the presentation of the automatic sequence controlled calculator by Mr. Thomas J. Watson, president of International Business Machines Corporation, and its acceptance by President James B. Conant, of Harvard University. The formal transfer of the machine will be attended by high ranking Navy officers, state and university officials, executives and engineers of International Business Machines Corporation, and representative leaders of science and industry.

The machine is completely new in principle, unlike any calculator previously built. An algebraic super-brain employing a unique automatic sequence control, it will solve practically any known problem in applied mathematics. When a problem is presented to the sequence control in coded tape form it will carry out solutions accurate to 23 significant figures, consulting logarithmic and other functional tables, lying in the machine or coded on tapes. Its powers are not strictly limited since its use will suggest further developments of the mechanisms incorporated.

In charge of the activity since the installation of the calculator in the Research Laboratory of Physics at Harvard is the inventor, Commander Howard H. Aiken, U.S.N.R. (B.S. University of Wisconsin '23, A.M. Harvard University '37, Ph.D. Harvard University '39) who worked out the theory which made the machine possible. Commander Aiken, now on leave as Associate Professor of Applied Mathematics in the Harvard Graduate School of Engineering, began his work on the device in 1935, when he joined the Harvard staff as Instructor in Physics and Communication Engineering.[2]

Two years of research were required to develop the basic theory. Six years of design, construction, and testing were necessary to transform Commander Aiken's original conception into a completed machine. This work was carried on at the Engineering Laboratory of the International Business Machines Corporation at Endicott, New York, under the joint direction of Commander Aiken and Clair D. Lake. They were

---

2. In place of this paragraph, the revised version has the following: "Commander Howard H. Aiken, now on leave as Associate Professor of Applied Mathematics at Harvard, originated the basic theory which led IBM to assign its engineers to invent a machine which would carry out the calculations desired by Aiken."

assisted in the detailed design of the machine by Frank E. Hamilton and Benjamin M. Durfee.[3]

Commander Aiken was assisted by Ensign Robert V.D. Campbell, U.S.N.R., during the latter years of the construction of the machine. Commander Aiken having been detailed to other Naval duties, Ensign Campbell, then a civilian attached to Harvard University, carried on in a liaison and research capacity.

The machine is of light-weight, trim appearance: a steel frame, 51 feet long and 8 feet high, a few inches in depth, bearing an interlocking panel of small gears, counters, switches, and control circuits. There are 500 miles of wire, 3,000,000 wire connections, 3,500 multiple relays with 35,000 contacts, 2,225 counters, 1,484 ten-pole switches, and tiers of 72 adding machines, each with 23 significant numbers.

When in operation in the soundproofed Computation Laboratory, the calculator is so light and so finely geared that it makes no more noise than a few typewriters.

The new calculator is not designed for a specific purpose, but is a generalized machine that will do virtually any mathematical problem. Among the many problems treated are: 1. Computation and tabulation of functions. 2. Evaluation of integrals. 3. Solution of ordinary differential equations. 4. Solution of simultaneous linear algebraic equations. 5. Harmonic analysis. 6. Statistical analysis.

*The remainder of the release (pages 4–8) describes the components of the machine, the way the machine solves problems, and the steps by which a mathematician puts a problem into a form in which it can be entered into the calculator. Two statements about the future are of interest. The first may indicate that Aiken really did envisage working on problems of orbit theory:*

When this calculator returns to civilian use and others like it are built, they will be of the greatest importance in astronomy in the

3. In place of this paragraph, the revised version has the following: "Over a period of six years, IBM's Clair D. Lake, pioneer inventor of printing accounting machine mechanisms, was in charge of the invention of the machine at the IBM laboratory at Endicott, N.Y., with Commander Aiken and IBM engineers James W. Bryce, Benjamin M. Durfee, and Frank E. Hamilton, all as co-inventors. This group conducted the development of the invention from beginning to completion of the machine. Commander Aiken began work on the theory of the device in 1935, when he joined the Harvard staff as Instructor in Physics and Communication Engineering. He has been assigned by the Navy to direct the use of the machine."

solutions of dynamic equations of the solar system, never solved because of their intricacy and the enormous time and manpower requirements.

*The second interesting statement, found in the closing sentences of the document, also deals with the future of the machine:*

It is already possible to visualize the peacetime functions of the calculator in pure and applied sciences. This machine is but the first step towards the establishment of a computation bureau which Commander Aiken hopes to establish for consultation by the research laboratories of science and industry.

# Appendix B
# Aiken's Talk at the Dedication

*The formal presentation of the Automatic Sequence Controlled Calculator to Harvard University was held in the Faculty Room of University Hall on Monday, 7 August 1944, at 2:30 P.M. Present were the Governor of Massachusetts, four admirals, various other officers of the armed forces, and invited faculty members, deans, and members of the Harvard governing boards, plus three of the IBM engineers who had constructed the machine and members of the operating staff of the machine. There were four talks in all, followed by a visit to the new machine. This appendix presents President James Bryant Conant's introduction of Howard Aiken and Aiken's address. The other two talks were by Dean Harald Westergaard and by Thomas J. Watson. Watson spoke about the needs of education.*

*The texts presented here are based on a stenotype transcript made during the actual presentations and show the signs of being talks rather than formal essays. The only editorial changes that have been made are some corrections of obvious errors in transcription and the occasional introduction of punctuation.*

*Aiken's two drafts of his talk are in the Harvard University Archives. Aiken, who did not like to read his talks, used his revised typescript as a kind of outline, reading some portions. The discussion that follows the transcript below notes some of the notable differences.*

## President Conant

Gentlemen, today Harvard welcomes a distinguished company and celebrates a unique occasion. We welcome here as our guests in Cambridge our friend Mr. Thomas J. Watson, and a group of high ranking naval officers, more particularly four officers of flag rank, Admiral Van Keuren, head of the Naval Research Laboratory; Admiral Furer, head of Co-ordination and Research of the Navy, with whom I have had the pleasure of sitting on the S.R.D. Advisory Council; Admiral Cochrane, head of the Bureau of Ships; and Admiral Theobald, Commandant of the First Naval District, who has assisted Harvard in so many different ways on so many different occasions. Indeed, one might almost regard this as a joint Navy-Harvard party, for reasons which will be evident shortly.

The unique occasion to which I refer is, of course, the formal acceptance of the Automatic Sequence Calculator, a gift of Mr. Thomas J. Watson, on behalf of the International Business Machines Corporation, of which he is the president. This marvelous calculating machine, which to me as a simple, bare-footed chemist seems as though it could do everything but talk—this machine is a joint development undertaken by a member of our Faculty of Engineering, Mr. H. H. Aiken,[1] and the engineers of the International Business Machines Corporation, in particular Mr. Bryce, Mr. Lake, Mr. Hamilton, and Mr. Durfee, all of whom, I think, are here today. This is a beautiful illustration of a co-operative endeavor brought to a successful conclusion.

The custodian of this gift is the Faculty of Engineering. For the time being, the machine is at full-time work for the nation. Under contract with the Navy, Harvard University is using this machine on important war problems presented to us by the Bureau of Ships and the Naval Research Laboratory. This is the reason why we have the privilege of having this group of high ranking naval officers with us today, to whom I have already referred. Commander Aiken, as he now is, on leave of absence from the Harvard Faculty, has been detailed here for the purpose of helping run this machine on the war problems.

I am going to ask two members of the Faculty of Engineering to speak this afternoon, first Commander Aiken, who, as I said, is in civilian clothes, Associate Professor of Applied Mathematics, Applied Mathematics in Civil Engineering, and he will describe the development of this machine and outline some of its potential possibilities for the future.

I shall then ask Dean Westergaard, Dean of the Faculty of Engineering, to say a word or two about the organization of the University and in particular the way in which that organization will fit into our plans for the use of this machine in the future.

In introducing our first speaker, Commander Aiken, I should like to take this opportunity to pay special tribute to his imagination, and

1. In those days, and for some decades later, there was at Harvard a kind of reverse snobbism in which a tenured member of the faculty (with the rank of Associate Professor or Professor) was called "Mr." The title of "Professor" was generally used only for non-tenured faculty members (with the rank of Assistant Professor).

the persistent skill in the development work, which, together with the techniques and skills of the engineers of the International Business Machines Corporation, and the constant support of Mr. Watson himself, for over a period of eight years, has make this remarkable development possible. Commander Aiken.

### Commander Howard H. Aiken, USNR

Mr. President and Gentlemen: I should state that the purpose for which we are forgathered here this afternoon is as old as civilization itself. Our purpose is to consider a device designed to assist in the solution of mathematical problems, and to derive numerical results. I say that purpose is old, because the record is clear that those who invented the fundamental processes of arithmetic were themselves the first to feel the need of mechanical aids.

The first mechanical aid to be used was, of course, the ten fingers of the hands. It is for that reason that the number system we still use at the present time is based on the numeral 10. After the invention of zero, and the extension of the numbers system, the fingers no longer sufficed for counting, and pebbles were used assembled in piles on sheets for the purpose. It was from this that the invention of the abacus came, wherein beads strung on wires took the place of the pebbles, and the wires facilitated their easy movement.

After the invention of the abacus the next step came at the hands of the man who produced the first multiplying device, known as Napier's number rods, or Napier's "Bones." And shortly after that— Kepler's invention of the use of the half-angle formula from trigonometry for multiplication, and later still Napier's logarithms were outstanding early achievements in speeding up computation.

The first calculating machine, apparently, was that of Pascal, who had in mind more the needs of his father's mercantile business than the scientific purposes for which one would have expected him to design such a machine.[2] And after Pascal came the work of Leibnitz, Newton,[3] Morland, Maxwell, and in fact nearly every great mathema-

2. Critical readers will note that Aiken erred in believing that Pascal's father needed a machine because of his mercantile activity, whereas he needed a machine for his accounts as a tax-collector.

3. Isaac Newton never designed or built a machine. Perhaps Aiken was thinking of Leibniz and Newton as inventors of the calculus.

tician of the early days who, at some time or another, turned his attention to the machine problem for speeding up the processes of arithmetic. All the early machine designers had in mind the design of a machine for the four fundamental processes of arithmetic—until the early part of the nineteenth century, when Charles Babbage in England for the first time tried to build a machine [that was] highly specialized and intended for the solution of scientific problems only.

Babbage spent nearly £50,000 of British money, obtained from the government on the recommendation of the Royal Society, in an attempt to build first what he termed a Difference Engine, and then what he called an Analytical Engine, after he had failed with his Difference Engine in the first place. I say Babbage failed, but I should like to make it especially clear that he failed because he lacked the machine tools and electric circuits and metal alloys, but through no fault of his own. Babbage's failure was due solely to one fact: he was a hundred years ahead of his time.

Babbage's failure and the distinct lack of tools and methods soon discouraged pure scientists and mathematicians in the calculating machine problem, with the result that the mathematician turned his energies toward other fields, and the problem lay for years untouched.

But after machine tools had been developed, and all the assets of modern manufacturing became theirs, the problem again was opened up by a variety of different manufacturers. Still, the problem remained in the field of the four processes, the four fundamental processes of arithmetic. And then there came one of the fundamental inventions of all times in the art of computation, the use of punched cards for the storing of numbers and for the rapid distribution of those numbers into counters for carrying on numerical processes. It was this invention, developed by the International Business Machines Corporation, and all the associated parts of mechanisms for speeding and using such cards, that has brought the possibilities of scientific calculating machinery again into a position where one could look at the situation with hopes of success.

This was the situation, then, as we saw it, as compared with the situation that Babbage saw one hundred and more years ago. It is now eight years since we began thinking in terms of automatic calculating machinery, especially designed for our purposes. And, after a preliminary period in which the mathematical theory alone of such machines was under consideration, a period in which the support of Professor Chaffee, Dean Westergaard, Professor Shapley aided this project, after that period we approached the International Business Machines Corporation and asked their support to build such a machine and construct it and put into operation.

Our first contact with that company was with Mr. J. W. Bryce. Unfortunately, his ill health has prevented his being with us here today, so I should like to take a few minutes and tell you a little about him. Mr. Bryce for over thirty years has been an inventor of calculating machine parts, and when I first met him he had to his credit over 400 fundamental inventions, something more than one a month. They involved counters, multiplying and dividing apparatus, and all of the other machines and parts which I have not the time to mention, which have become components of the Automatic Sequence Controlled Calculator that you are to see this afternoon.

With this vast experience in the field of calculating machinery, our suggestion for a scientific machine was quickly taken and quickly developed. Mr. Bryce at once recognized the possibilities. He at once fostered and encouraged this project, and the multiplying and dividing unit included in the machine is designed by him. I have already commented on Mr. Bryce's health, and this fact alone prevented his constant association with the work throughout the six years that followed, and left us in a position where we were free to have his advice, but unfortunately not his constant association throughout the job.

On Mr. Bryce's recommendation, the construction and design of the machine were placed in the hands of Mr. C. D. Lake, at Endicott, and Mr. Lake called into the job Mr. Frank Hamilton, and Mr. Benjamin M. Durfee, two of his associates, and there began a cooperative enterprise between the four of us which has not yet ended, even though the machine has been finished.

The early days of the job consisted largely of conversations, conversations in which I set forth requirements of the machine for scientific purposes, and in which the other gentlemen set forth the properties of the various machines which they had developed, which they had invented, and based on those conversations the work proceeded until the final form of the machine came into being.

One of the greatest disappointments I suffered during the time of this work was that, because of the war, I was called into active duty in the Navy, and for a long period was only remotely associated with the machine in the latter stages of its construction. During that time Ensign Campbell of the United States Naval Reserve took my place and carried on until the job had ended. I can't say too much in praise of those who have worked on this construction. All have contributed enormously.

It remains only for me to point out first that this project, the construction of this machine, differs very greatly from other relations between industries and universities in the past. I do not know that this calculating machine is the first example of the precedent that has been set here, but it is well nigh the first, if not really the first. In the past, the lines of communication between industries and universities have been more or less one-way lines of communication. Universities have supported research projects, and these have been aided by industry in the form of grants of money and gifts of apparatus, and in the maintenance of fellowships. And on the other hand, members of the faculty of the university have gone out into industry and have assisted industry in the solution of technical problems or in other ways as needed.

But here has been a project in which representatives of an industry and a university have worked side by side with a common end in view in collaboration. To me this is almost as important as the mechanism itself, because I believe that the day when a physicist or an engineer, working alone for production and research, while that day is not over, it is near its end. The tools of research have become larger, and the time has come when the physicists need methods of a production engineer in many of the projects he engages in. And so I feel that the co-operative phase of this development and the precedent it is setting is one which should be well borne in mind in the future.

I have commented on all those who have worked on this project and mentioned the parts which they have played. I have yet to mention Captain Solberg and Commander Ferrer, who, hearing of the machine's completion and delivery to Harvard, at once took action in the Navy Department to see that the machine was put to work on the solution of problems in the interest of the national defense. It is for that reason that I have been sent back to the University to carry on with the machine.

And now—when the facts were presented to the International Business Machines Corporation back some six years ago, the data available was in the form of equations. I described how this machine was to operate, with the crudest of sketches suggesting the different types of machine components that might be used. There were tables of numbers representing constants that should be included in the processes, but that intangible result of thought process was all that existed. And it was on these intangible things that Mr. Watson made his decision to support this project and foster it and bring it to a conclusion. Surely, under what was offered, there were few men who would have expended the resources of their company, their men and their money, for this work, and certainly no one could have done more. Thank you.

*Aiken's talk more or less followed the same line of development as the draft version. Of the eleven pages of the text or draft, about three are devoted to the prehistory and early history of machine calculation. As in the spoken version, there is praise for Babbage and an explanation of his failure (which is attributed to the low state of technology at that time). Aiken omitted from his talk the following paragraph concerning Babbage's successors:*

> Babbage's researches stimulated other mathematicians to design and construct difference engines of smaller capacity which ultimately were completed and employed in the computation of mathematical functions. These machines were of limited capacity and equipped with a limited number of difference wheels. Those who after Babbage succeeded in their efforts were the Scheutz brothers [actually, father and son], and Mr. George Grant of Cambridge, Massachusetts. In his paper in the American Journal of Science, Mr. Grant acknowledges the help and encouragement of Professors Eustis, Winlock, and Whitney of Harvard College, thereby establishing an early precedent in the University in the design of calculating machines.

*In the talk, Aiken described the first stages of his invention somewhat differently than in the text. In the text he said that eight years earlier, when he had first contemplated his dream machine, the central problem was "the design of automatic sequence control mechanisms by means of which standard machine elements could be made to carry out the arithmetical processes automatically without the supervision of an operator." The eventual solution was to use "methods based on the calculus of finite differences" so that "machine elements could be controlled so as to obtain the solutions of differential equations. and of definite integrals and be made to tabulate functions, perform harmonic analyses, and carry out almost all of the mathematical processes of applied mathematics."*

*In the draft, but not in the spoken version, Aiken referred to the future of the machine after the war. "It is the plan of the University," he said, "to employ the machine as the central element of the computation laboratory designed to carry out different problems of interest not only for scientists of the University, but for other institutions as well." He envisaged applications to problems of astronomy, physics, and engineering, statistics, life insurance, and also economics "and other topics far removed from the physical sciences."*

*Aiken knew that his listeners would want to have some measure of the speed of the new machine. In the draft, he said that the only true way to evaluate the speed would be first to solve a complex problem by "manual methods" and then* to do so by machine. Aiken confessed that he was "not sufficiently interested in the subject to expend 2 or 3 months for no other reason than to find out the speed with which the calculator operates." However, in setting up a demonstration for the visitors, Aiken had found, "for the first time," a problem sufficiently simple that "an actual estimate of the of the ratio of machine to manual methods could be made"—namely, "the computation of the Bessel functions $J_0$, $J_1$ through $J_6$ in the interval from 2 to 4, by increments of one hundredth, with an overall accuracy of 15 places of decimals." Aiken confessed that the ratio of time for machine solution versus time for manual solution was "so optimistic" that he felt a need to check "on the speed of manual computation" with two senior members of the faculty who were expert in such calculations: Edwin C. Kemble of the Physics Department and Bart J. Bok of the Harvard College Observatory. The conclusion was that the new machine was "well nigh 100 times as fast as a well equipped manual computer." Additionally, in considering the two relative speeds, it had to be borne in mind that a human computer "can work little more than 6 hours a day before fatigue causes him to produce a prohibitive number of errors." The "sequence controlled calculator," Aiken concluded, "operating on a 24 hours a day schedule," was able to "produce as much as 6 months' work in a single day."

Apart from the variations just noted, the greatest difference between the two versions is in the discussion of IBM's role in designing the new machine. Furthermore, in the draft, but not in the spoken version, there is a paragraph acknowledging the assistance and support of Harvard officials: "Professors Chaffee, Mimno, Shapley, Brown and Mr. Cunningham, all of whom encouraged the development in the early days when the theoretical background of the machine was under consideration." The draft also mentioned Dean Westergaard, who "took great interest in the machine and sponsored it as a project of the Graduate School of Engineering."

In the draft there are only two sentences about Bryce, but in the spoken version three paragraphs are devoted to Bryce and his contribution to the project. There is a real difference between the two versions in Aiken's presentation of the contributions of Lake, Hamilton, and Durfee. In the draft, Aiken said that these three "engaged in conversations designed to present the advantages of different types of calculating machine elements and to bring about the adjustments necessary to meet the requirements of a machine for the solution of scientific problems." Also, "These early conversations, nearly six years ago, formed the basis of the design necessary to construction." The spoken version enlarges considerably on the contributions of this trio, evidently reflecting Aiken's reaction to Watson's anger at the Harvard news release's portrayal of Aiken as the inventor of the machine.

In the draft Aiken referred to his being called to active duty and to "*my research assistant, Mr. Robert Campbell, now Ensign Robert Campbell, USNR, [who] took over my job until the machine was completed and delivered to the University.*" He also gave credit to "*Mr. Lee Smith and Mr. Wesly Pfaff,*" who "*have constructed and assembled most of the special mechanisms employed in the machine and have been associated with its development almost from the beginning.*" So far as I know, these two IBM employees have never been mentioned in histories of IBM or of the ASCC/Mark I.

Aiken's draft also mentioned that the "*standard parts of IBM manufacture and all of the special mechanisms have been produced in the company's factory by many different employees, most of whom are unknown to us. Their help and labor must also be acknowledged, especially in view of the care and accuracy with which the fabrication of the parts has been carried out.*"

The final paragraph of the draft, an encomium of Thomas J. Watson, is much the same as in the spoken version.

# Appendix C
# Aiken's Memorandum Describing the Harvard Computation Laboratory

*This memorandum, dated 1949, is of great interest for a number of reasons. First, it recounts what Aiken believed to be the significant accomplishments of the Computation Laboratory during its early years. Second, it shows Aiken's recognition of the fact that agencies of the federal government were acquiring their own computational facilities and his awareness of the consequences of this change. Third, it indicates the central problem of financing Aiken's computer activities. Fourth, today's readers will be impressed by the extremely small amounts of money that were of such concern in 1945. Fifth, Aiken sets forth clearly a rule (later adopted universally at Harvard University) that "classified" research—that is, research whose results cannot be made generally public—is inconsistent with the nature of a university. Sixth, Aiken stresses the need for the development of computer logic as well as componentry.*

The staff of the Computation Laboratory began operations with the Automatic Sequence Controlled Calculator as a Naval activity under the auspices of the Bureau of Ships, May 15, 1944. The Calculator, usually referred to as Mark I, was constructed by the International Business Machines Corporation at Endicott, New York, according to the joint designs of the present writer and engineers of the company. After completion, it was removed to Cambridge, installed in the basement of the Research Laboratory of Physics, and presented to the University by the International Business Machines Corporation on August 7, 1944.

On February 1, 1945, the Bureau of Ordnance of the Navy Department entered into a contract with Harvard University under which the staff of the Computation Laboratory began research leading to the construction of the Mark II Calculator. Later, this contract was amended to include the operation of Mark I when this facility was released by the Bureau of Ships at the close of the war. As work on Mark II progressed, it became clear that the space available in the basement of the Research Laboratory of Physics was totally inadequate for the work required by the Government. Accordingly, arrangements were completed whereby the rental received from Mark I operations could be used to amortize a large part of the cost of the present Computation Laboratory building.

Mark II Calculator was completed and delivered to the Naval Proving Ground at Dahlgren, Virginia, during the month of March, 1948. This machine has proven very satisfactory for the engineering calculations required at that activity, the latest report being that the machine delivered useful results 92% of the operating time during the month of August this year. Further, the operating record continues to improve as time goes on.

Before Mark II was completed, the Bureau of Ordnance proposed further amendments to the contract, calling for the construction of the Mark III Calculator, at present under final test here at Harvard. It is expected that this machine will be delivered to the Navy sometime prior to March 1, 1950. During the construction of this machine, engineers representing a number of foreign governments and industrial concerns have visited the Computation Laboratory for periods ranging from a few months to a year, as set forth in the following tabulation.

*Table I*

| Agency | No. of Representatives | Period |
|---|---|---|
| Sweden | 3 | 1 year |
| Switzerland | 2 | 6 months |
| Belgium | 2 | 1 year |
| India | 1 | 3 months |
| Holland | 1 | 3 months |
| Czechoslovakia | 2 | 2 months |
| General Electric Company | 2 | 4 months |
| Engineering Research Associates | 1 | 2 months |
| Telecommunications Laboratories | 2 | 3 months |
| Teleregister Company | 4 | numerous visits |

When Mark II Calculator was finished, the Bureau of Ordnance no longer required the computational services provided by Mark I. Hence, the machine was made available to the Air Force and to the Atomic Energy Commission under whose joint sponsorship it has operated during the present year. Negotiations are at present under way leading to the extension of these contracts to cover operating costs during the year 1950.

In 1947, it became increasingly clear to the present writer that the intensive development of computing devices in progress here at Harvard and elsewhere should be accompanied by a corresponding educational program in Applied Mathematics with Special Reference to Computing Machinery. Accordingly, the Office of Naval Research was asked to sponsor a one-year program of instruction leading to the degree of Master of Science. The program was first offered during the academic year 1947–48. The following year, the program was again offered under a contract with the Air Force. In all, seventy-six students have enrolled in this program, at least part-time. Fourteen M.S. degrees and one Ph.D. have been given.

When representatives of the Air Force contracted for the operation of the Mark I Calculator and for instruction in Applied Mathematics, they also agreed to support a research program in computing machine components. This progress has led to the development of the magnetic delay line and the development of a circuit synthesis, both of which have attracted considerable attention.

From the foregoing discussion, it should now be clear that the Computation Laboratory has been supported almost exclusively by Federal Funds allocated under a number of contracts as set forth in Table II.

*Table II*

| Agency | Contract | Amount |
| --- | --- | --- |
| Bureau of Ships | NObs-14966 | $ 14,800.00 |
| Bureau of Ordnance | Nord-8555 & Nord-10449 | 1,520,871.90 |
| Office of Naval Research | N5ori-76 | 25,817.25 |
| Air Force | W19–122-ac-24 | 257,685.00 |
| Atomic Energy Commission | AT-(30–1)-487 | 41,096.00 |
| Total | | $1,860,270.15 |

This government support cannot, however, be expected to continue indefinitely since:

(1) As the government obtains its own computing facilities it will no longer need the computational services provided by Mark I. This was made clear by the action of the Bureau of Ordnance when Mark II was delivered.

(2) The development, both here at Harvard and elsewhere, of better computing machinery has made Mark I almost obsolete. Hence, it will soon be difficult to support the machine on any basis.

(3) At the recent "Second Symposium on Large-Scale Digital Calculating Machinery" the present writer announced that he would not undertake the construction of any further calculators excepting the possible construction of a Mark IV to be installed in the Computation Laboratory at Harvard University. The reason for this decision is that sufficient work has already been done to enable industry to take over such manufacturing operations, thus leaving the laboratory staff free for component and other research.

(4) The present writer is unwilling to undertake "Classified" work in the Computation Laboratory since such work is in large part inconsistent with the very purpose for which the University exists. This position has been respected by both the Navy and the Air Force. However, renewal of the Atomic Energy Commission contract is at present at doubt because of security questions.

From the foregoing comments it should be apparent that private support must be given the Computation Laboratory in the very near future if it is to survive.

The best evidence in favor of such support is, of course, the results thus far attained. To this end, the accomplishments of the staff of the Computation Laboratory to date have been summarized in Table III.

Since August 1944, when Mark I was completed,[1] interest in digital computing machines has increased immensely. For example, 336 people attended the first "Symposium on Large-Scale Digital Calculating Machinery," January 7 through 10, 1947, on the occasion of the opening of the present Computation Laboratory building, while 701 attended the "Second Symposium" September 13 through 16, 1949, on the occasion of the announcement of Mark III Calculator. . . .

1. Aiken errs here. Mark I was completed and ran its first problem in Endicott in January 1943. It was moved to Harvard and became operational in March 1944. In August 1944, IBM formally presented the machine to Harvard.

As the initial constructional phase in the activity of the Computation Laboratory draws to a close, it becomes necessary to examine the objectives which should be attained in the near future. These include:

(1) Construction or procurement of a Mark IV Calculator for installation here at Harvard. This machine should be available for the solution of problems arising in the University community. It should not be leased to some outside agency as has been the practice in the past, thereby making the machine unavailable to the Harvard Faculties. In order that as little time be spent on this objective as possible, Mark IV should be highly similar to Mark III.

(2) Increased activity in applied mathematics, development of numerical methods, and collaborative solution of problems arising in the research conducted throughout the University. The recent report, "Double Refraction of Flow and the Dimensions of Large Asymmetrical Molecules," by Scheraga (Cornell University), Edsall (Harvard Medical School), and Gadd (Computation Laboratory) is an excellent example.

(3) Research in computing machine components, logical design of machines, coding systems, etc. All this work should be directed toward smaller, more effective computing devices of low cost. Actual construction should be left to industry.

(4) Research in automatic control devices. Most of the techniques employed in automatic calculators find application in a variety of control problems. The solution of these is largely logical rather than experimental. When the Theory of Communication and the Theory of Information are brought to bear on control problems, their solutions fall within the scope of the Computation Laboratory.

(5) Most important of all, greater emphasis should be placed in both undergraduate and graduate instruction in applied mathematics.

The needs of the Computation Laboratory include:

| | |
|---|---|
| Mark IV Calculator | $300,000 |
| Purchase of tools and furnishings | 100,000 |
| TOTAL EXPENDITURE | $400,000 |

In addition, the annual cost of operation is approximately $100,000 of which less than $3,000 is at present supplied by the budget of the Department of Engineering Sciences and Applied Physics, the remainder being supplied by the Government.

As regards the construction of Mark IV Calculator to be permanently installed in the Computation Laboratory, little further comment should be necessary. Without a machine the staff of the Laboratory is reduced to operating obsolete equipment.

The item "Purchase of Tools and Furnishings—$100,000" is made necessary by the fact that all of the tools and furniture in the Computation Laboratory building are the property of the Government. It is not impossible that the Navy may decide to leave this equipment at the Laboratory after contract work has been completed, in which case this item may be saved.

The annual operating cost of the Laboratory, $100,000, includes $75,000 for calculator operation and $25,000 for component and other research. Since these items have been the subject of discussion on several occasions, they are known to be completely realistic by various members of the faculty and other officers of the University.

It remains only to observe that the Computation Laboratory, because of the basic nature of numerical procedures, cuts horizontally across the vertical departmental organization of the University. It is not unusual to find the coding staff engaged in work on problems arising in the fields of physics, economics, statistics, and engineering while the component and design groups are dealing with problems of electronics and mechanical engineering—all at the same time. As the Theory of Communication and the Theory of Information develop new methods for extending the digital computer's technique to such problems as, for example, industrial control, air traffic control, and the mechanical translation of languages, so will this relatively new field of applied research bring new benefits to society.

Cambridge, October 19, 1949
Howard H. Aiken
Director of the Computation Laboratory

# Appendix D
# The Stored Program and the Binary Number System

Stanley Gill has called the stored program the "key design feature" of modern computers. As the name suggests, the central property of the stored program is that instructions are "held" in the computer's internal storage "while they are awaiting execution." In the early machines—such as those of Aiken, Stibitz, and Zuse—all instructions were fed into the machine as a sequence of operations coded onto punched cards, paper tape, or film. These stored commands have been described as "nothing more than a form of read-only memory containing the program instructions." Mark I had three such readers, and according to Gill there was "provision for control to be passed from one to another, to allow some flexibility in the logical structure of the program." Because, in such a system, the instructions had to be read seriatim at a speed determined by the mechanical features of the tape advancement mechanism, this method of providing instructions was much too slow to be used in high-speed electronic machines, such as ENIAC. The designers of ENIAC, therefore, used a kind of plugboard programming, of the sort originally developed for punched-card tabulators. While this method of programming was clumsy and laborious, it must be kept in mind that ENIAC was originally designed for a limited set of problems and was not planned to have its program changed frequently. As is well known, however, Eckert and Mauchly and those associated with them in the original design of ENIAC had been thinking ahead to the general problem of flexible controls.[1]

1. One of the hotly debated questions in the history of computers is who invented the concept of the stored program. It is now agreed that the concept was developed by the designers of ENIAC and first put into a coherent statement (with an original twist) by John von Neumann. Machines based on this concept have come to be known as "von Neumann machines."

Once electronic machines, ENIAC's successors, were developed, designers had to face the problem that "no tape reader could scan instructions fast enough to keep up with the internal speed of the computer." Accordingly, internal storage of instructions became a feature of the new breed of electronic machines. This innovation made possible a rapid succession of innovations, introducing a number of features which vastly increased the flexibility and capabilities of the computer and which collectively have become known as the "stored program." It is probably not an exaggeration to say that the computer as we know it today came fully into being with the advent of the stored program.

Brian Randell, in a lengthy letter to me, summed up the main features of the stored program as follows.

First of all, there was the "recognition that a program can itself be viewed as data and hence could result itself from earlier computation." Charles Babbage, in the middle part of the nineteenth century had a sense of this aspect of the stored program in the "formula cards" which his proposed machine was to develop. Babbage "even wrote of his realization that by ensuring that his Engine would be able to produce cards as output there would be no need for a special machine (which he had for a while considered) for partially automating the task or preparing programs, and reducing the incidence of coding errors in them." Mark I lacked this feature of Babbage's proposed design and "was not capable of punching sequence control tapes."

Second, "programs must be capable of determining (based on the results of calculations to date) what operations should be applied to what operands and when." This feature implies the ability "to select data dynamically (typically by somehow calculating data addresses) as well as being able to select between alternative next instructions, and being able to repeat instruction sequences—all in ways that are influenced by prior calculations." In the "absence of something like an indexing or an indirection technique (or Turing's scheme of being able to use the contents of a register as the next instruction to be executed), the ability to actually modify a stored program gives one (by rather convoluted programming) the required flexibility of operand selection, and of subroutine calling and return." It is important, however, to note that "if one does have the necessary facilities in the instruction code, then it is indeed a good idea to avoid altering the program that is being executed." In the sense just set forth, Mark I clearly had no facility for program modification. It is often claimed that Mark I was later altered so as to have the capability to perform conditional branching. In actual practice, however, as Randell takes note, "all that was provided was a means of signalling to the operator that a condition had occurred." That is, "Mark I provided 'mere' condition testing, and relied on operators to select and position sequence tapes manually, in order to complete the provision of conditional branching."

Third, instructions "should be held in the same (or the same sort of) store as the data," a feature that "facilitates assuring that instructions can be made available at a rate commensurate with data availability—a crucial issue for electronic computers." One of the reasons why "early advocates urged use of a single actual common store (for variables, data, and instructions) was that different applications needed different amounts of apace for these three different sorts of information." There is no evidence that Aiken, at least early in his career in computers, appreciated this aspect of the stored program.

Aiken, throughout his career, insisted on maintaining the separate identity of instructions and data. As a result, he obviously could have no sympathy with a concept that would permit the machine to alter a program. On this score, Randell notes, "It is true that the practice of having executing programs alter themselves is now frowned upon." As a result, "Aiken's views on this issue are indeed in line with contemporary thought." It is important, however, according to Randell, to keep in mind that "until instruction codes provided the sort of flexibility described above, such alteration had some important practical benefits." It is now recognized, furthermore, that "one does not have to go to the extremes of having separate stores in order to ensure that such alteration does not happen." One "can make sure that programs are not altered, even if they are held in the same store as data, by the use of a storage protection facility."

Aiken's rejection of the stored program in his architecture for his four machines had consequences that rather completely offset any proposed advantage in maintaining the "integrity" of programs and data. The Aiken architecture deprived his machines of the capability of reading instructions at a high speed commensurate with the speed of the machine. Another disadvantage was that there could not be random access to any part of the program which would be needed as the work progressed. That is, with a stored program the computer could jump almost instantaneously from one point of the stored program to another without having to scan a program tape which would entail considerable loss of time. There is no doubt in the minds of pioneers and historians that in the early days of the electronic computer the stored program exercised a major role in making possible the enormously rapid advances in the computer with which we have become familiar.[2]

One further feature of the stored program needs to be stressed. Computer designers quickly became aware of yet another advantage that followed on having the instructions put into the same word length as the data. This feature, with its implication that both instructions and data would be kept in a single storage unit, not only simplified computer design, but also influenced the mode of operation. For example, in computers having this feature, the program would have the same accessibility as the data. As a result, common storage of instructions and data quickly became features of computers.

Aiken's Mark IV, however, his final venture into electronic computers, did not have such common storage. Rather, Mark IV was designed to have separate storage for instructions and data, reflecting Aiken's philosophy of maintaining the integrity of data and instructions. It followed as a consequence of this conservative point of view that Aiken never approved of the computer manipulating or altering the program in the same fashion as the data.

Aiken was clearly taking a position counter to the mainstream of computer development. Perhaps this was a factor of his decision, announced while Mark IV was under construction, that he would build no more computers. In the 1973 interview, however, Aiken insisted that the reason why he gave up the design and production of such machines was quite different. By this time, he said, industry had taken over and was building computers. In his opinion, industry was better suited and equipped for this purpose than universities. In accord with this philosophy, Aiken arranged for the purchase by the Harvard Computation Laboratory of an industry-produced computer, a UNIVAC, which was of course a stored-program machine.

Aiken's prejudices concerning some features of the stored program seemed reactionary to his contemporaries and must seem reactionary today. Certainly, the designers and builders of the new generation of computers in the late 1940s and the 1950s—EDSAC, EDVAC, BINAC,

2. Mark III had separate magnetic drums for storing data and instructions. Mark IV had an ingenious system of internal storage, still keeping data and instructions in separate stores. Though it improved the accessing of parts of the program and the data, the use of separate storage units did not have the advantages of the stored program.

IAS, UNIVACs, and others—would have seen Aiken as a conservative, a holdout against progress.

In the long run, however, as has been mentioned, one feature of Aiken's reactionary stance has become today's standard practice. The maintenance of the integrity of the program has become a significant feature of much of our present computer design. Gill notes, in this regard, that although "alteration of programs during execution enormously increased the scope of automatic computing" and as such was "heavily used in the early days," its use has "diminished considerably." Among the developments which "demanded the abandonment of program alteration during execution," Gill cited "the use of read-only memories for programs needed very frequently," an "approach now rather widely used in microcomputers."

When we were planning the celebration of Aiken and his four machines for Pioneer Day in 1982, Bob Campbell brought to our attention the fact that the Texas Instruments Company, a leading manufacturer of computers, had been advertising machines featuring the unalterability of programs by referring to their design as based on "Harvard architecture." If I read their promotional literature correctly, a more appropriate name would have been "Aiken architecture."

Aiken took a very conservative stand on pure binary arithmetic as well as on the stored program. That is, he expressed opposition to the exclusive use of binary arithmetic in computer design. This subject came up during the 1973 interview. His point of view seemed puzzling because binary arithmetic makes possible the high speed of present computers. Furthermore, all of nature, life, and thought seems to exhibit binary or dual properties: on and off, plus and minus, male and female, good and bad, open and closed, living and dead, black and white, light and heavy, sun and moon, and so on. Additionally, simple logic—such as used in computers—is binary or two-valued.

I gathered from Aiken's remarks that one of his points was that computer designers had become so "fixed" on binary that they no longer even considered the possibilities of other systems. He was opposed to what seemed a kind of intellectual straitjacket. After all, he explained, the numerical input is always in decimal, as is the desired output. Accordingly, the computer has to have two "extra" sequences, one to convert the input from decimal to binary and the other to convert the output from binary back into decimal. Aiken's position does not have so firm a basis as he believed, however, since—as shown by von Neumann and others (e.g. Herman Goldstine)—the conversion from one number system to another is relatively simple and requires very little machine space and machine time.

In this connection it must be kept in mind that there are many different ways of representing decimal digits by groups of binary digits and we may well understand that Aiken believed that a better understanding of this whole subject would be profitable. In a personal communication, Maurice Wilkes recalled his discussions with Aiken and reminded me of Aiken's belief that "when this subject was better understood, it might be possible to choose a binary representation for the decimal digits that would make it possible to design decimal computing circuits that would be competitive with binary circuits." Aiken devoted much time and thought (and much time on Mark I) to this study and published some of his results in volume 27 of the Annals of the Harvard Computation Laboratory, *Synthesis of Electronic Computing and Control Circuits* (1951). Binary representations for decimal digits were of great interest to Wilkes, who recalled that he "studied the book closely when it came out."

# *Appendix E*
# *Aiken's Three Later Machines*

In November of 1944, soon after Mark I began producing useful work for the Navy, the Navy Bureau of Ordinance commissioned Harvard to design and build a second machine, to be installed at the Naval Proving Ground at Dahlgren, Virginia. This machine, designated Mark II, was designed and assembled by Aiken and his staff. The naval authorities, Aiken noted, "urged that the machine be so designed that its construction could be completed as soon as possible." The need for this calculator was so pressing that there was no thought of experimenting with wholly new or untried devices or circuit elements. Hence, "it was decided to build a relay calculator," very similar to Mark I, but with improvements suggested by the actual use of Mark I.

The Navy and Aiken, furthermore, were very satisfied with the operation of Mark I and the results it was continuously producing. It must be remembered that at this time, so far as actual computing machines were concerned, Mark I represented the state of the art. ENIAC was still under construction and would not become operational until the war's end. Since Mark I did not have the capacity to serve all the Navy's needs, the Navy wanted a second computing machine that would be similar to Mark I and that would do equally good work. Among some major improvements deriving from the actual experience of using Mark I were better relays and the use of 10 rather than 23 digits.[1]

1. Today we know that at that time, unknown to Aiken or the Navy, there were three computer developments (at least) that rivaled Mark I. One was the series of electronic machines built and operated at Bletchley Park, in London, for code-breaking. Another was the series of machines designed by Konrad Zuse in Germany (notably the Z-3). The third was the ENIAC, the electronic machine that would eventually make Mark I obsolete. The series of relay

The new plug-in relays were designed by the staff of the Computation Laboratory "in collaboration with Harold Seaton, electrical engineer of the Autocall Company, Shelby, Ohio," which manufactured them. For obvious reasons, Aiken decided to use as few IBM components as possible. So, instead of the IBM input and output devices found in Mark I, Mark II had tape punches, tape readers, and page printers supplied by the Western Union Telegraph Company. These were adapted from standard teletype printing equipment according to the specifications of the Comp Lab staff, so as "to meet the requirements of digital computing machinery."

Before undertaking the Navy's assignment for a second machine, the project had to gain the approval of IBM for both moral and legal reasons. Therefore, as soon as Aiken received instructions to "design . . . a controlled sequence calculator for the Navy," President James B. Conant of Harvard wrote a letter to Thomas J. Watson Sr. (dated 26 December 1944) informing him of the Navy's plans. After stressing that "the design and procurement of this calculator is regarded as urgent and necessary to the war effort," Conant got to the heart of the matter:

We do not know of any reason why we should not furnish the facilities requested. However, in view of our past relations with you, we are bringing the matter to your attention and ask that you advise us whether you can see any objection to our complying with the Navy's request.

Conant concluded with an appeal to Watson's vanity and his patriotism:

The request seems to be another proof of the great importance of the undertaking which you and your company so farsightedly sponsored. The production for the United States Government of a useful new instrument of war in time of crisis like this is something on which all who are concerned with the development of such enterprises may well be proud.

Evidently Watson raised no objections, and Aiken and company set out to plan and build Mark II while still operating Mark I on a full schedule.

One important difference between Mark II and Mark I was the elimination of the mechanical feature of rotating bus or cam shafts with

---

machines (starting from the original one designed for complex-number arithmetic by George Stibitz) built at the Bell Telephone Laboratories were neither so powerful nor so versatile as Mark I.

clutches to advance ten-position counters. Furthermore, Mark II was a double machine, consisting essentially of two parts, each of which could be operated separately or as part of a single whole. Mark II speeded up the computing process by storing numbers in two-position relays rather than in ten-position registers or counters as in Mark I. Since the two positions of such a relay correspond to "on" and "off" (0 and 1), four relays were required to represent a decimal digit using the binary system, in which the four relays have assigned to them: respectively, 8, 4, 2, and 1. Representing each numerical quantity in a calculation required 46 relays.

Like Mark I, Mark II was programmed through a sequence of numerically coded commands on a punched paper tape. Constants were entered in a series of ten-position switches, as in Mark I. Furthermore, for each of the two parts there were built-in special functions and, of course, a set of interpolators. Mark II was thus much more flexible than Mark I and operated much more quickly. Table E.1 gives comparative operating times.

Each half of Mark II had one addition unit, two multiply units, and six built-in function units (for $X^{-1}$, $X^{-1/2}$, $\log X$, $\exp_{10}X$, $\cos X$, and $\arctan X$) in addition to electromechanical relay storage registers, special functional registers, and controls for interpolations. Each half had also its

*Table E.1*
Operating times (seconds).

| Operation | Mark I | Mark II |
|---|---|---|
| Addition | 0.3 | 0.2 |
| Multiplication | 6.0 | 0.7 |
| Reading tape | | 1.5 |
| Positioning interpolator and reading values from tape | | 8.0 |
| Division | 11.4 | 4.7 |
| Reciprocal square root | | 6.0 |
| $\log_{10}x$ | 68.4 | 5.2 |
| $10^x$ | 61.2 | 6.7 |
| $\cos x/\sin x$ | 60.0 | 7.5 |
| $\arctan x$ | | 9.5 |
| Punching tape | | 1.5 |
| Printing | | 4.0 |

own numerical data punched-paper tape reader, interpolator, data-tape-punching mechanism, teleprinter, and instruction-tape-punching mechanism. Both machines (or both parts of the machine) were served by the same manual (or keyboard) unit for preparing instruction tapes and the same manual (or keyboard) unit for preparing data tapes. Unlike Mark I, which was strictly decimal with a fixed point, Mark II used floating-point data words. Mark II had a data-word format that made use of decimal digits represented in binary coded decimals, while the instruction word format was in both octal (external) and binary coded octal (internal).

Mark III was very different from its two predecessors. In 1948, when the architecture of this machine was under consideration, ENIAC and been in operation for two years, proving that a rapid all-electronic calculating machine could indeed function with an unexpectedly high degree of reliability. Initial research and conceptual planning for Mark III, also intended for Dahlgren, began in 1946, while Mark II was being assembled. The actual detailed design and construction was not undertaken, however, until the summer of 1948, soon after Mark II had been sent off. In March 1951, after assembly and testing at Harvard, Mark III was moved to Dahlgren and started regular operation in January 1952. Mark III ran until 1955, when it and Mark II were retired some time after the arrival of an IBM machine named NORC.

Mark III differed architecturally from its two predecessors. It made use of magnetic drums for storage, and it had both an electronic adder and an electronic multiplier. Formally named Electronic Magnetic Drum Calculator, it was described in the Manual of Operation as "an electronic computing machine in which numerical quantities and coded operating instructions are stored with the aid of magnetic pulse recording techniques."

While there were some electro-mechanical relays, there were also 4500 vacuum tubes, some of which (in miniature) were designed and made especially for Mark III. Some were triodes and dual triodes; others were pentodes and twin diodes. Additionally, the circuits of Mark III contained germanium crystal diodes. Mark III did not have any provision for entering the program in the same storage as the data and did not make use of the stored-program principle.

One of the novel features of Mark III was that numbers and instructions ("in binary-coded decimal form") were stored in this machine on nine aluminum drums, coated with "a thin film composed of finely

*Table E.2*
Decimal digits: 2*, 4, 2, 1 notation.

|   | 2* | 4 | 2 | 1 |
|---|----|---|---|---|
| 0 | 0 | 0 | 0 | 0 |
| 1 | 0 | 0 | 0 | 1 |
| 2 | 0 | 0 | 1 | 0 |
| 3 | 0 | 0 | 1 | 1 |
| 4 | 0 | 1 | 0 | 0 |
| 5 | 1 | 0 | 1 | 1 |
| 6 | 1 | 1 | 0 | 0 |
| 7 | 1 | 1 | 0 | 1 |
| 8 | 1 | 1 | 1 | 0 |
| 9 | 1 | 1 | 1 | 1 |

divided magnetic oxides of iron suspended in a plastic lacquer, and applied to the drums with an artist's air brush." These drums rotated at about 9900 rpm. It was observed that such "magnetic drum storage has the pleasing characteristic that information once recorded remains on the drum indefinitely, unless deliberately altered." Furthermore, changing a binary digit from 0 to 1 or from 1 to 0 did not require first erasing the existing digit by demagnetization.

Eight tape drives were provided for reading and recording information on oxide-coated paper tape. The data-word format consisted of 17 decimal digits with adjustable decimal point, in which each decimal digit was in a hybrid or modified binary code. The control and data input tapes were magnetic. The magnetic tape and magnetic drum units and the electronic circuits were all designed and built at Harvard by the staff of Computation Laboratory.

In storing numerical information by using only the binary digits 0 and 1, Aiken did not use the ordinary binary system, as he had in Mark II; he introduced a hybrid system in which there were four binary digits with "weights" or values 2*, 4, 2, 1. This system is illustrated in table E.2. Here "2*" serves to distinguish the value 2 in the fourth place from the 2 in the second place. The Manual of Operation stated that the system adopted for Mark III had these desirable features:

- Examination of the 1 component alone indicates whether a digit is odd or even.

- Examination of the 2* component alone indicates whether a digit is less than 5, or 5 or greater.
- The nines complement of each decimal digit is obtained by inverting the binary digits, 0 and 1.
- Three of the four binary components have the same weights as in the binary number system, permitting many of the simple properties of this system to be retained.

Each cycle on Mark III required about 4 milliseconds. Hence, addition was several orders of magnitude faster than on Mark I or Mark II. Multiplication, requiring 3 cycles, took 12 milliseconds, but printing was still slow, requiring 3.5 seconds for a 16-decimal-digit word. While Mark III did useful work, there were a number of problems of unreliability. A major cause of this troublesome feature was traced to the cooling off and warming up when the machine was shut down for weekends and then tuned on again. This kind of problem was eliminated by allowing it to run continuously.

In 1950, at a conference held in Cambridge, England, on "High-Speed Automatic Calculating Machines," W. S. Elliott reported that, according to Professor Aiken, Mark III was "the slowest electronic machine in the world," requiring "12½ milliseconds for multiplications." Nevertheless, Mark III "will invert a matrix of order 50 in 90 minutes." Mark II is "300 or 400 times faster than the Mark I calculator and six men are kept busy preparing problems for the Mark I." He quoted Aiken as having said that "if all the machines now being built are completed, there would not be enough mathematicians to run them." He then reported Aiken's prediction that Mark IV would be even slower than Mark III. Elliott then remarked that Aiken "emphasized" that the "speed of getting errors out of a machine may be determined much more by the frequency of occurrence of errors than by the inherent operating speed."[2] He concluded by referring to Aiken's views on the need "for easing the work of mathematicians in coding," a point on which Aiken "feels strongly." There is "only one way to lick coding," according to Aiken, and "that is to eliminate it." Accordingly, Mark III was being equipped with a "mathematical but-

---

2. Elliot then described some calculations that Aiken had made. If the "inherent operating speed" is increased by 10, while the error rate remains the same, then there is "very little improvement in overall speed." But if the "inherent speed" is kept constant and the rate of error is reduced by a factor of 10, then "the overall speed is multiplied by 3."

ton board." By its means, mathematicians will be given a means of coding Mark III automatically, since the "button board . . . understands normal mathematical language" and "prepares all the necessary orders," even calling up "the sub-routines" for the machine.

Mark IV differed from its predecessors in many ways. For one thing, Mark I, Mark II, and Mark III were associated with the Navy, but Mark IV was designed and built at Harvard under a contract with the Air Force. Mark II and Mark III were removed from Harvard to Dahlgren when completed, but Mark IV remained at Harvard. The history of Mark IV goes back to early 1948, when Aiken signed a contract with the Air Force. The contract had two main provisions. First, courses were to be developed to train specialists in mathematics, with primary attention to automatic computing machines. Second, an investigation was to be undertaken of mathematical and physical techniques that could be used in a computing machine. Fifteen students were sent to Aiken by the Air Force for courses beginning in the fall of 1948. By 1950 the design and construction of Mark IV were underway. The machine was completed in 1952. John Harr, a member of the design team for Mark IV, has noted that the main innovation in Mark IV was "the first random-access memory, employing magnetic shift registers, a technique developed at Harvard." Mark IV was all-electronic, using electronic (vacuum) tubes that were chiefly miniature, with glass envelopes. Selenium diodes were used in the rectifier board in the arithmetic unit, but these were later replaced by germanium diodes. In addition to the large magnetic drum, there were more than 30,000 magnetic cores.

Mark IV was, in many ways, similar in its architecture to Mark III. The drum system for Mark IV "included a consolidation of all storage on one drum, new thinner multiple-channel magnetic read/write heads," and new circuitry which introduced a greater reliability than had been the experience of Mark III. This drum memory could store 4000 16-bit decimal-digit data words and 10,000 20-bit instructions. The most striking difference between Mark IV and its predecessors was the complete absence of relays. Both arithmetic and control functions "were primarily performed with diode logic." One of the features of Mark IV was that the instruction preparation unit "used a keyboard that allowed a user to enter instructions in an algebraic notation and then automatically generated and wrote machine code on a magnetic tape." This feature had first been tried on Mark III. Mark IV was Aiken's last machine and the first to be all-electronic. Mark IV was not a stored-program machine.

In Mark IV, Aiken did not use the new and rapid electrostatic storage devices because they were still plagued by unreliability. Instead, Mark IV had a "random-access memory" consisting of "200 magnetic-core shift registers." There was also "bulk storage in the form of a large magnetic drum with separate sections for 400 16-digit data words and for 10,000 instructions." All "arithmetic and control functions were performed by selenium rectifiers, later replaced by more reliable germanium diodes."

Mark II was a great favorite of computer programmers and operators because of its property of being a twin or dual machine. It functioned well during its seven years at Dahlgren. Mark III was retired after some five years. Mark IV had a long and useful life in the Harvard Comp Lab, serving not only the Air Force but also outside users and a generation of students.

# *Appendix F*
# *How Many Computers Are Needed?*[1]

Howard Aiken is often quoted as having declared that only a very small number of computers would be needed to serve the needs of the whole world—perhaps a dozen, with eight or ten for the United States. Sometimes the number is given as six or even two or three. Documentary evidence confirms that Aiken did indeed once say that one or two "computers" would suffice, but he does not seem to have been thinking of all the possible uses of computer power. The context shows that his remark did not have the general meaning that might be supposed and that it was not, therefore, as outrageous as it might appear.

An early report on Aiken's prediction concerning the number of computers (the first one I have found in print) appears in the record of an interview with J. Presper Eckert and John Mauchly published in the April 1962 issue of the journal *Datamation* ("An Interview with Eckert & Mauchley," *Datamation* 8 (1962), no. 4: 25–30). The interviewer, Harold Bergstein (editor of *Datamation*), asked "Who made the now classic prediction that the total requirement for computing power in the U. S. would be six machines?" Eckert replied that there was "some confusion" on this topic, and that the confusion reflected a misinterpretation of something Mauchly had once said, which was "Let's get orders for six machines to get us enough backlog to make it worthwhile to go ahead with the ENIAC project." Then, according to Eckert, "Howard Aiken made a remark that if six machines were built [we would] never be able to train enough people to program enough problems for these machines." Eckert observed that he didn't "know" whether Aiken "was referring to the six machines we were

---

1. This appendix is derived in large part from I. Bernard Cohen, "Howard Aiken on the Number of Computers Needed for the Nation," *IEEE Annals of the History of Computing* 20 (1998), no. 3: 27–33.

building or whether he was referring to some other source." Eckert remarked, in conclusion, that Aiken "was thinking in terms of just hand programming" and didn't "take automatic programming into account." These comments show that Eckert was aware of Aiken's argument about the limited number of mathematicians available for programming.

In the ensuing discussion, Mauchly said he believed that Aiken hadn't taken "into account the fact that some of these machines would be devoted to repetitive tasks such as running payrolls weekly, for which you can keep these machines busy without large programming efforts producing new problems all the time." Eckert expressed the opinion that Aiken had not "thought of sorting and collating and file maintenance and all that sort of thing which take up hours and hours of time on these machines [when] you're not really doing any computing in the ordinary sense of the word."

None of the three men who participated in the above-mentioned dialogue—Eckert, Mauchly, Bergstein—seems to have been aware of an earlier prediction by Aiken, recorded in some notes taken by Edward Cannon (John Curtiss's assistant at the time), that only two or three large-scale computing machines would be needed. Cannon apparently did not make public his notes about Aiken's prediction until 1973, when he introduced this information into his testimony in the famous court case of *Honeywell v. Sperry Rand*.

When Aiken made the above-mentioned prediction, in March 1949, no computing machines that made use of the stored-program concept were in operation. Aiken made the prediction after a meeting of a subcommittee appointed by the National Research Council of the National Academy of Sciences. This group, officially called the Subcommittee on High Speed Computing and also known as Subcommittee Z, was chaired by John von Neumann.

The specific immediate assignment of the subcommittee was to evaluate three computer projects under consideration by the National Bureau of Standards: one submitted by the Electronic Control Company of Philadelphia (organized by J. Presper Eckert and John Mauchly), one by the Raytheon Manufacturing Company, and one by the Moore School of Engineering at the University of Pennsylvania. The only prospective user of such a computer ("high-speed computing machine") named in the subcommittee's report was the Bureau of the Census, which had been interested in the work of Eckert and Mauchly and in the possibility of using a post-ENIAC machine. The Bureau of the Census, however, was not permitted to fund development work; for this reason, all arrangements for a possible new computer were made through the Bureau of Standards. The Bureau of the Census wanted an appraisal by experts of the practicality of an Eckert-Mauchly machine and assistance on specifications to be made in any contractual arrangement if the proposal was a sound one.

John H. Curtiss of the Bureau of Standards was pleased with the proposed assignment to Subcommittee Z. Aware of the needs of his own bureau in this domain, he saw the action of this subcommittee as an opportunity to strengthen the bureau's position as the central patron in the development of the new computing machines for various government agencies. In 1946, $300,000 had been transferred from the Army Ordnance Department to the Bureau of Standards for the development of a computer. Before long, bids and proposals were solicited and three rival systems were put into competition: one from the Moore School, where Eckert and Mauchly had been on staff during the first discussions with the Bureau of the Census; one from the new company formed by Eckert and Mauchly, who had left the Moore School; and one from Raytheon.

Thus, when the Bureau of Standards turned to the National Research Council for technical advice, there were three rival proposals to be evaluated. According to Nancy Stern, the historian of UNIVAC, the bureau hoped that in the final report of Subcommittee Z some comparative information would also be made available concerning computers associated with the subcommittee's members, among whom were George Stibitz (formerly of the Bell Telephone Laboratories), John von Neumann (Institute for Advanced Study), Samuel Caldwell (MIT), and Howard Aiken (Harvard). The machines in question were Bell Laboratories' series of relay machines, the IAS machine and MIT's Whirlwind (both then under construction), and the Harvard machines. This list thus includes at least seven machines (or families of machines), five of which were being planned to be stored-program computers. An important part of Subcommittee Z's assignment was to consider not merely the three rival proposals of specific machines but also the plan to order five new stored-program computers.

However, in a draft of the subcommittee's report, dated 16 March 1948, the only machines under discussion are the one proposed by Eckert and Mauchly, the one proposed by Raytheon, and the one proposed by the Moore School, all three of which were planned to be computers in some important senses in which we use that term today.

From the viewpoint of the Bureau of Standards, the report was apparently unsatisfactory because of the absence of any comparative evaluations of the IAS machine, the Harvard machines, or Whirlwind. The subcommittee also reported on various policy issues concerning the activities of the Bureau of Standards in developing computers.

The subcommittee met on 9 March 1948 in Cambridge.[2] All four members were present, as was John Curtiss. The report (written by von Neumann) noted that many design features of the three proposed machines were basically the same. Accordingly, these three machines, in the words of the subcommittee's report, did not represent "three really different and independent intellectual risks." A "choice between the three proposals, on a primarily technical basis" was "hardly possible," since the progress reports of the three rival groups did "not show to what extent the essential componentry that is being described has been developed, submitted to life tests, and found reliable under the conditions which the computing machine would have to use it." Nor was there any real information concerning "realistic estimates of the time and the expenditures involved."

The final part of the report is especially significant in view of Aiken's remark about the need for such computers. It was noted that the "Bureau has about $900,000 available to build five computing machines." Thus, "the amount available for one computing machine is $180,000." But the estimates given by the three "proposed contractors" quoted "prices for machines based on the Bureau's specifications" which were as high as "$650,000 per machine."

The report clearly and unmistakably concludes that fewer than five machines should be built, and that the contractors should go back to their drawing boards. The committee found the overall specification of the machines too "general purpose" for the specific needs of the Bureau of the Census.

The subcommittee's report stated that, in view of the complexity of the planned "computing machines," the first step should be to produce a set of working components rather than a complete machine. The report stressed that "the design and construction of large-scale digital computing machines represents an art which is still in the research and development stage." Accordingly, it was "unwise at this time" to

---

2. The date and the place are mentioned in the draft report. I do not know whether the subcommittee had held earlier meetings, nor whether there were later meetings.

build a series of "duplicate calculating machines," and "even more unwise" to "initiate the building of a higher number of identical ones." After working components were developed and fully tested, a decision should be made as to whether the new machine or machines should be assembled by the bureau or by one of the three constructors.

In a 1978 interview conducted by Nancy Stern, Isaac Auerbach remembered that the committee concluded that "it was impossible to design and build a mercury memory that would operate in a computer." At the time, Auerbach stressed, he and Brad Sheppard had "designed, built, and demonstrated such a memory." Auerbach "never saw the report," but he "recall[ed] [their] surprise when [they] heard about [the] conclusion." "All they had to do," he noted, "was get on a train, and 40 miles away in Philadelphia they could see a computer mercury memory operating." The committee should have stopped "conjecturing about the theoretical aspect of things"—they should have "just come down and look[ed]."

The staff of the Bureau of Standards apparently favored the proposal of Eckert and Mauchly. Indeed, it is a fact of record that the bureau did not follow the subcommittee's recommendations and in the end proceeded to support Eckert and Mauchly in the development of the UNIVAC series. This partiality is in strong contrast, as Nancy Stern has documented in her book *From ENIAC to UNIVAC,* to the subcommittee's evident bias against Eckert and Mauchly.[3] This generally negative feeling, Stern found, could be traced back to the days of World War II, when the ENIAC project was not highly regarded by the experts; they considered Eckert and Mauchly and their group "naive and unsophisticated." Both Caldwell and Stibitz had displayed negativity toward the ENIAC group in their official positions during World War II. Stern points out that von Neumann, even though he had been associated with Eckert and Mauchly during the days of the ENIAC, had become "particularly unsympathetic to Eckert and Mauchly" and "regarded the commercial interests of these two men as unprofessional."

Explaining why von Neumann and the committee had not bothered to find out what Eckert and Mauchly and their associates were doing, Auerbach referred to "the von Neumann—Eckert and Mauchly feud, which eventually broke up what was to become a joint venture." He

3. In 1948, Aiken did not have a high regard for Eckert and Mauchly and their new company. See chapter 1 above.

traced the origins of the "major rupture" to von Neumann's "effectively [having] indicated that he was going to control the conceptual design of the next machine." Aiken, Auerbach continued, considered Eckert and Mauchly "the enemy," because "they had one-upped him." He had been "one-upped by ENIAC, which kind of stole his thunder." Because Aiken had an "extraordinary ego," Auerbach concluded, he "just could not tolerate anybody not considering him *the* most eminent computer engineer in the world." "Von Neumann also had a supreme ego," Auerbach added. The "only other person" in that ego league "would be Norbert Wiener," but he was "a more modest man by comparison."[4]

This hostility toward Eckert and Mauchly, when coupled with the general attitude expressed in Subcommittee Z's report, provides a context for understanding Aiken's apparently gloomy prediction about the number of computers. Aiken's statement was made at the time of the meeting of the subcommittee, but not during the formal sessions. It was recorded by Edward Cannon. According to Cannon's memorandum, Aiken "expressed to the two of us[5] in a rather pontifical manner that we were misleading the government and public" by "trying to provide for the development of large computers of the type we were interested in." The "reason" Aiken gave, according to Cannon, was that "there never would be enough work for more than two of these computers." Accordingly, "rather ought we to turn to assisting in the development of small, desk-sized computers." Cannon noted that this was not a "conversation" with Aiken; a "sermon by Dr. Aiken is the best description."

Aiken was recorded to have declared, furthermore, that "we" (that is, Subcommittee Z) would be "misleading not only the [government] agencies which had made money available for development of such equipment, but the general public in pursuing the course of trying to develop such equipment, giving the impression that this program could be justified." The reason, Aiken is said to have argued, was "that there will never be enough problems, enough work for more than one or two of these computers." Accordingly, Aiken concluded by telling

4. The magnitude of Auerbach's categorization can be fully appreciated only by those who had personal contact with Norbert Wiener, for whom "modest" would be the least appropriate adjective in the English language.

5. It is not clear whether Aiken's remark was made to Cannon and Curtiss or to Cannon and Samuel Alexander.

Cannon and either John Curtiss or Samuel Alexander (and the National Bureau of Standards) "to go back and change your program entirely"—to "stop" this "foolishness with Eckert and Mauchly." The Cannon memorandum, with its quotation of Aiken's opinion, assumed historical importance when Cannon read it into his testimony in the celebrated case of *Honeywell v. Sperry Rand* (p. 17,935).[6]

The bluntness of Aiken's reported language and Cannon's description of his style ("pontifical manner," "sermon") testify to the accuracy of the report. In order to understand what Aiken had in mind, one must take into account that he was talking in the context of the meeting of Subcommittee Z. The purpose of the subcommittee was to deal with the new breed of digital stored-program machines—actual "computers" in very nearly the currently accepted sense of that term. Aiken certainly knew of at least two such large-scale stored-program machines in the process of being designed and built in the United States: John von Neumann's "IAS machine" at the Institute for Advanced Study and Jay Forrester's Whirlwind at MIT. Two members of the subcommittee (von Neumann and Caldwell) were associated, respectively, with these two machines.

An analysis of Aiken's actual words and a consideration of the context help to make Aiken's meaning clear. Aiken said that the Bureau of Standards was "misleading the government and [the] public" by its activity with regard to the "large computers of the type we were interested in." This would indicate that Aiken was discussing the specific types of machine under consideration by the Bureau of Standards, not large-scale computing machines in general. Recall that the bureau's proposal called for five such machines. Aiken said—and these are his very words as recorded by Cannon—that "there never would be enough work for more than two of these computers." Therefore, it seems clear that Aiken was not making a prediction about the work to be done in every aspect of computer work in the United States, but was referring specifically to the proposed computers ("these computers") that would be used by the Bureau of the Census or possibly the Bureau of Standards. Two of the new breed of computers would suffice for these bureaus, Aiken concluded. Accordingly, "rather ought we to turn to assisting in the development of small, desk-sized

---

6. I do not know whether Aiken's remark was made to Cannon after the Cambridge meeting of the subcommittee or ar some later meeting. But the context would have been the same at all meetings.

computers." In retrospect, then, this first part of Aiken's recorded statement is in no sense completely absurd, as it would seem to be when quoted out of context and as if it applied to the total computing needs of the nation.

The interpretation I have just given is actually reinforced by Aiken's next remark: that "we" would be "misleading" the government and the public by "giving the impression" that "this program" (that is, the program of building five of the new kind of computers) "could be justified." In support of his position, Aiken argued that "there will never be enough problems, enough work for more than one or two of these computers." As has just been mentioned, Aiken's reference to "these" computers was apparently to the computers to be built for the Bureau of the Census and the Bureau of Standards, not the number of such computers for every imaginable purpose in the United States. Therefore, his advice to the staff of the bureau was "Go back and change your program entirely" and to "stop" the "foolishness with Eckert and Mauchly."

One part of Aiken's reported statement to Cannon is of special interest: his prediction that there might not be "enough problems, enough work" to keep new high-speed machines busy. At the time of the meeting of Subcommittee Z, there seems to have been a somewhat widespread feeling that computers of this new breed might work so quickly that they would soon run out of problems to keep them busy. Evidence for the prevalence of such a belief can be found in a talk on "The Future of High-Speed Computing" that John von Neumann gave in November 1949 at an IBM Seminar on Scientific Computing. Von Neumann began by taking note of a "major concern which is frequently voiced in connection with very fast computing machines." There was a fear, he said, that, because of "the very high speeds which may now be hoped for," these machines "will do themselves out of business." That is, the new machines "will out-run the planning and coding which they require and, therefore, run out of work." Accordingly, although von Neumann did not say so specifically, the implied reaction to this fear would be that it would be unwise to build many such machines.

Von Neumann's own position on this question is of great interest. He began by taking note that, for "problems of those sizes which in the past—and even in the nearest past—have been the normal ones for computing machines," the actual "planning and coding required much more time than the actual solution of the problem, even on one of the hoped-for extremely fast future machines." But, he pointed out, "the problem-size was dictated by the speed of the computing machines then available." Von Neumann believed that an equilibrium would eventually be reached when the problem time became relatively longer ("but not prohibitively longer") than the "planning and coding time." In other words, new problems of greater complexity were being addressed by the machines. With the new fast machines, von Neumann argued, there would be the same kind of pressure, acting like an "automatic mechanism"—a "pressure toward problems of larger size." He envisaged that there would be "a year or two, perhaps, during which extremely fast machines will have to be used relatively inefficiently while we are finding the right type and size [of] problems for them."

Von Neumann obviously did not anticipate the advent of programming languages that would radically alter the ratio between planning and coding time and problem-solving time. In the present context, the significant point of von Neumann's rebuttal is that he was fully aware of an envisaged limitation in the potential use of the future computers—of the predicted possibility that they might "run out of work." But von Neumann also knew that the argument for needing only a few machines was based on the need to solve problems on the scale of and of the sort then being solved on existing machines. Von Neumann was aware that the new generation of high-speed, electronic, stored-program computers, such as those being constructed at the Institute for Advanced Study and at MIT and those being considered for the National Bureau of Standards, would produce a radical change in the type, size, and scale of problems. That is, ranges of problems that had once been ruled out of consideration because of their complexity and scale would be brought to the computer.

While Aiken stressed the paucity of mathematicians, he was not quite so bearish on the future of high-speed computers. The feature story about computers that appeared in the issue of *Time* dated 23 January 1950 was, in large measure, based on an interview with Aiken that must have taken place late in 1949, only about a year after Aiken made the remarks Edward Cannon recorded. The article took note of the increasing speeds in the sequence of Aiken's machines. Mark II was said to be "ten times as fast as Bessie [i.e., Mark I]," and Mark III to be "25 times as fast as Mark II." The "machines now building" would be "faster still." Evidently, Aiken was asked what the consequences of this increased speed would be. He is quoted as having said "We'll have to think up bigger problems if we want to keep them busy." This statement would seem to reinforce the interpretation that Aiken's response to the advent of high-speed computers was not that one or two of them would suffice for all the computing needs of the United States.

In 1952, four years after Cannon recorded Aiken's remarks and three years after von Neumann's presentation to the IBM Seminar on Scientific Computing, Aiken discussed the change in the type of problems for which "computing machines" were used. The occasion was a pair of talks, one given at Fairleigh Dickinson College and one at the Harvard Graduate School of Business Administration.[7] Speaking six years after ENIAC became operational, Aiken contrasted the perceived goal and actual assignment of the early "calculating machines" and those of the new computers. At first, he recalled, "there was no thought in mind that computing machines should be used for anything except out-and-out mathematics." Aiken stressed this feature of the early machines: "I do not exaggerate when I say that everyone in any way connected with the original development of large-scale computers had in mind only one thing—namely, the construction of machines for the solution of *scientific* problems." "No one was more surprised than I," Aiken admitted, "when I found out that these machines were ultimately to be used for control in business." He and others had designed and constructed machines to solve differential equations or systems of differential equations or to produce tables of functions. These problems had come from engineering, the natural sciences, and—to some degree—the social sciences and statistics. "Originally one thought," Aiken reminisced, "that if there were a half dozen large computers in this country, hidden away in research laboratories, this would take care of all requirements we had throughout the country."[8] This is the only reference I have ever found to Aiken's having specifically referred (in a public talk, a manuscript text, a letter, or an edited statement) to the need for only a few large-scale "computing machines." The phrase "originally one thought" would seem to indicate that this point of view was rather commonly held but does not tell us specifically whether Aiken was thinking along these lines.

In both the talk at Fairleigh Dickinson and the one at the Harvard Business School, Aiken went on to explain that computers were no

7. Partly edited transcripts of both of these talks are in the Aiken files of the Harvard University Archives. Parts of the texts are reprinted in *Makin' Numbers.*).

8. This sentence occurs in the Fairleigh Dickinson transcript but not in the Harvard Business School version.

longer restricted to the solution of mathematical problems. Vast new areas of application had been discovered: varieties of data processing (including commercial billing and control), statistical analyses and investigations of new sorts, varieties of automation in business and in manufacturing, machine translation from foreign languages, and so on. In other words, the new computers were not limited to the original domains of science and engineering for which the first "computing machines" had been designed, but were being applied to a constantly expanding series of new uses.[9]

In retrospect, it would seem that Aiken's remark to Cannon was in part an expression of his personal disdain for Eckert and Mauchly. But, as the foregoing analysis shows, predictions about the number of computers should be read in context and not displayed as general statements about the all the future computer needs of the country. Aiken did not deny the fecundity of his own brainchild.

9. During the 1973 interview, I asked Aiken about the statement of the need for only a limited number of computers. He began to answer, referring to von Neumann's ideas; then, as so often happens in interviews, the subject shifted and we never got back to it.

# Appendix G
# The NSF Computer Tree

One sign of the importance given to Mark I in the early years of the computer is a widely disseminated "family tree" in which Aiken's machine holds a prominent place. This tree occurs in a number of variations. The original version was designed by the National Science Foundation and was displayed at an international conference held in 1959 in Paris under the sponsorship of UNESCO.[1] In this diagram, the main trunk is labeled HARVARD MARK I. Branches stemming from the main trunk bear leaves labeled BELL LABS MODELS I—VI, ZUSE, MARK II, MARK III, MARK IV, MONROBOT, and C.E.C.E. The upper trunk of the tree, labeled ENIAC, splits into IAS TYPE and EDVAC. The branches and leaves of the spreading tree are then labeled to illustrate the proliferation of computers in the 1950s.

This presentation is open to criticism as a misrepresentation of the historical evolution of the computer. For example, it indicates that Zuse's machines were a small twig branching out from the main trunk of Mark I and that ENIAC was developed from Mark I, whereas in fact both came into being independent of Mark I. Hence, it is easy to understand the adverse critical reaction this tree must have aroused.

Konrad Zuse's autobiography contains an account of his experience at "the Paris UNESCO conference," where he saw what he calls "an inaccurate table of the computer, which showed that to a certain extent my work stemmed from Aiken's work." Zuse was quite right; the tree indeed does show his work as a small branch growing out of Mark I. Zuse says that he at once "lodged a formal protest with the representative for German Affairs at the conference," but "nothing came of it." I assume that Zuse meant either that the poster was not taken down

1. I have not been able to determine whether the NSF tree was displayed as a poster, distributed as a souvenir, or both.

or that distribution of the "tree" was not canceled at once. The German official to whom Zuse referred was Alvin Walther, a professor from Darmstadt who was a longtime friend and admirer of Aiken. Walther would have recognized at once that his countryman's objections were legitimate, at least insofar as the tree gave an incorrect representation of the relation between Zuse's machines and Mark I.

No doubt in response to criticism such as Zuse's, a somewhat altered tree was issued by the National Science Foundation in 1960 in a pamphlet titled *"The Family Tree" of Computer Design: A Brief Summary of Computer Development.* Here it is said that the "tree" was produced by the NSF and displayed at "Automath '59," an exhibition in Paris in conjunction with the First International Congress of Information Processing, sponsored by UNESCO. This tree shows the same overall pattern as the former one, with the same labeling of the trunk, the roots, and the branches. However, a comparison leaves little doubt that the tree in the pamphlet is a later and revised version.

In the revised tree, Mark I and the four roots are marked off from the rest of the tree by a field of white dots. The upper trunk and the branches are divided into two groups: "parallel" computers are now indicated by a shading of black stripes, with letters in black, "serial" machines by solid black with white letters. Braces at the right divide the machines into four classes: "roots" (the four main roots), "predecessors," "first generation" (including the Bell Labs machines, Zuse's, Mark II, and Mark III), and "present generation" (all the rest). The main trunk, including the Harvard Mark I, seems to straddle "predecessors" and "first generation."

In the modified tree, Mark I is the only machine in the category of "predecessors." This tree—which still gives the misleading impression that Zuse's machines were an offshoot of Mark I—was given wide circulation when it was reproduced in Gordon Bell and Alan Newell's 1971 book *Computer Structures: Readings and Examples.*

There are many later versions of the family tree. In one, which may be found in the library of the Charles Babbage Institute, there is no designation for the roots. In this diagram, "Harvard Mark I 1944" has been moved downward from the main trunk into a giant urn from which the main trunk (now designated ENIAC) springs. In some later versions, all the roots are eliminated and there is no reference to Mark I.

# Appendix H
# Who Invented the Computer? Was Mark I a Computer?

In assessing Howard Aiken's role in history, three interlocking questions often arise: What is a computer? Was the ASCC/Mark I a computer? Who invented the computer? The answers to the second and the third of these questions depend on the definition that emerges from the first.

Historians of technology and computer scientists interested in history have adopted a number of qualifications that define a computer. As a result, the question of whether Mark I was or was not a computer depends not on a general consensus but rather on the particular definition that is adopted. Often, some primary defining characteristics of a computer are that it must (1) be electronic, (2) be digital (rather than analog), (3) be programmed, (4) be able to perform the four elementary operations (addition, subtraction, multiplication, and division) and—often—extract roots or obtain information from built-in tables, and (5) incorporate the principle of the stored program. A machine does not generally qualify as a computer unless it has some further properties, for example the ability to perform certain specified operations automatically in a controlled and predetermined sequence. For some historians and computer scientists, a machine must also have been actually constructed and then become fully operational.

The first machine put into operational service as a computer that embodied the stored-program concept, was digital and electronic, and had an adequate memory was the EDSAC, designed and constructed at Cambridge University under the leadership of Maurice Wilkes. As Wilkes acknowledges, the EDSAC "was designed according to the principles expounded by J. Presper Eckert and John W. Mauchly, and others at the summer school held in 1946 at the Moore School of Electrical Engineering in Philadelphia." The EDSAC ran its first calculation on 6 May 1949.

The design of the EDSAC was strongly influenced by the celebrated report, originating from the Moore School of the University of Pennsylvania, setting forth the principles of a machine known as EDVAC, an all-electronic digital machine planned to embody the stored-program concept. Historians have long debated whether this landmark document was the brainchild of John von Neumann or whether it represented ideas developed by the Moore School group, primarily John Mauchly and J. Presper Eckert. Although the "conceptual design" of EDVAC was complete in 1946, the machine itself did not become fully operational until 1951. As a result, EDVAC was not the first actual stored-program machine to be built. It is often held that the first stored-program machine to be built was the one designed and constructed at the University of Manchester by F. C. Williams and T. Kilburn. This, however, was a very limited device, intended primarily to test the Williams tube memory.

The literature on the history of computers is filled with rival claims for the honor of having been the "first" computer. But most historians of technology are not primarily interested in "firsts." For one thing, as Michael Williams has said, "If you add enough adjectives to something it will always be first." Claims about "first" seem to be based on a false idea of how history unfolds. It seems to be assumed that "in the beginning" there was a perfect Platonic ideal of the final or eventual computer, and that various inventors or applied scientists were in a race to fulfill it. But history shows that there was no such race. Rather, there was a series of innovations and improvements. There were also fruitless moves, each of which at one time seemed to point to the next development.[1]

Sometimes it has been said that there was a race between Maurice Wilkes at Cambridge and Tom Kilburn (under the direction of F. C. Williams) at Manchester. Wilkes reports, however, that Kilburn told him that if there was a race it was "to establish the viability of the Williams tube storage before [Max] Newman spent the money on importing an American (designed) machine."

I am concerned less with priority than with the properties for which various machines have been said to have been computers. A review of these properties will help to answer the question of whether Mark I was a computer.

---

1. Two examples of the latter are the Josephson junction and the magnetic bubble.

Some writers would give place of honor to Alan Turing, who conceived a computer-like machine when no such machine had ever been designed and built. Others would give credit to Charles Babbage, whose proposed Analytical Engine—although mechanical and not electrical or electronic—had a number of basic features found in modern computers.

Among other pre-EDSAC machines that may have been computers, a high place would certainly be given to a series of machines built in Britain during World War II and used to break codes. The later machines in this series bore the name Colossus. In 1996 a working replica of Colossus was built by Tony Sale and installed in a museum created at Bletchley Park, where the original machines did valiant service. In an announcement of a lecture by Sale at the Science Museum in South Kensington, a claim was made that Colossus was "the World's first working electronic computer." This claim was discussed at length in an article in the London *Financial Times* for 27 and 28 July 1996. At issue was whether Colossus or ENIAC was the "first" computer. What may be of most interest in the present context is that there was no question but that both machines *were* computers in a currently recognized sense.[2]

On any scale of innovation, great honors must be assigned to the machines invented by Konrad Zuse in Germany. In many ways, Zuse seems to have been the first person to conceive or invent a machine that meets many of the qualifications for having been a computer. In honoring Zuse, we would also take note of the important contributions made by his collaborator Helmut Schreyer.

Konrad Zuse told me an amusing anecdote about how he first encountered the work of Aiken. The occasion of our conversation was a luncheon in Zuse's honor, hosted by Ralph Gomory at the Watson Research Laboratory of IBM before a lecture given by Zuse to the staff of the lab. When Zuse learned that I was gathering materials for a book on Aiken, he told me that he had first come across Aiken and Mark I in an indirect manner, through the daughter of his bookkeeper. She was working for the German Geheimdienst (Secret Service) and knew through her father of Zuse's work on a large-scale calculator. According to Zuse, the young woman never learned any details about

2. Although it "had no stored program and was set up before each run by means of plugs and switches on a board," and although it "had virtually no memory," Colossus was—according to Sale—a very early example of parallel processing. "Separate logic calculations were being carried out on each of the five longitudinal tapes."

his machine, which was shrouded in wartime secrecy. But she knew enough about Zuse's machine to recognize that the material filed in a certain drawer related to a device that seemed somewhat like Zuse's. She reported this event to her father, giving the file number of the drawer, and the father at once informed Zuse of her discovery.

Zuse, of course, could not go to the Secret Service and ask for the document since that would give away the illegal source of his information. Zuse was well connected, however, and was able to send two of his assistants to the Secret Service, armed with an official demand for information from the Air Ministry, requesting any information that might be in the files concerning a device or machine in any way similar to Zuse's.

Zuse's assistants were at first informed that no such material existed in the files, but they persisted and eventually got to the right drawer. There they found a newspaper clipping (most likely from a Swiss newspaper), containing a picture of Mark I and a brief description about Aiken and the new machine. But there was not enough technical information to enable Zuse to learn the machine's architecture.[3]

Two other important pre-World War II computer pioneers were John Atanasoff[4] (a professor of physics at Iowa State University) and George Stibitz (an applied mathematician working at the Bell Telephone Laboratories). Atanasoff claimed to have introduced the designation "analogue computer"; Stibitz seems to have been responsible for the term "digital computer."

In 1939–40, Atanasoff, wanting a machine that would solve simultaneous equations, built a working prototype of a machine that used vacuum tubes and binary arithmetic.[5] Stibitz's initial motive was his desire for a device that would automatically add complex numbers or vectors. His ideas led to the development of a series of important relay calculators built during and after World War II.

3. See Konrad Zuse, *The Computer—My Life* (Springer-Verlag, 1991), p. 127. During the 1973 interview, Aiken recalled encountering Zuse at a meeting in Germany and remarked on Zuse's strong feeling that he had not received worldwide credit for his innovations. Among the Aiken papers in the Harvard University Archives is a small piece of 35-mm movie film punched for use as a program tape in one of Zuse's machines, evidently a souvenir given to Aiken by Zuse.

4. Atanasoff achieved a measure of fame in 1973 during a patent dispute between Sperry-Rand and Honeywell; Judge Earl Larson, in his decision, declared Atanasoff "the inventor" of the computer.

5. Atanasoff worked with a graduate student named Clifford Berry.

Another machine sometimes said to have been the first computer is IBM's SSEC, which was intended to surpass the ASCC/Mark I and thus to outdo the achievement of Howard Aiken. The SSEC had vacuum tubes, an electro-mechanical memory, and some branching capabilities. A number of historians concur with Paul Ceruzzi's judgement that it was "a true computer."

Many historians and computer scientists hail ENIAC—built by John Mauchly, J. Presper Eckert, and their associates at the Moore School—as a computer and even as the first computer, even though it did not embody the concept of the stored program. For example, Emerson Pugh, in *Building IBM*, asserts without qualification that ENIAC was "the world's first large-scale, programmed, electronic computer." According to Maurice Wilkes, ENIAC was "a real computer"; indeed, it was "the first large-scale electronic computer" in "the sense that it was an arithmetical machine"; it "just wasn't a stored-program computer."

In 1996 a number of celebrations were held to mark the fiftieth anniversary of ENIAC and what was considered to be the advent of the computer. Vice-President Al Gore stated flatly that ENIAC was "the first computer in the world," and in 1997 President Bill Clinton referred to the machine (though somewhat obliquely) in his second inaugural address.

"Never before in the history of computers," Arthur and Alice Burks say of ENIAC, "had an invention produced this order of magnitude of progress." Mark I and other early machines were able to do the work equivalent to that of several dozen human operators with desk calculators. But ENIAC was so many orders of magnitude faster and more powerful, especially with regard to its internal calculation speed, that it is even now hard to make meaningful comparisons with the machines that preceded it. In a lecture given at Harvard's 1947 Symposium on Large-Scale Digital Calculating Machinery, Lewis P. Tabor of the Moore School gave a few examples of ENIAC's speed. Formerly, he said, "a set of firing tables required a staff of twenty-five, equipped with desk calculators, about three months." ENIAC, with a staff of five, was able "to accomplish the same result in about one day." Tabor then astonished the audience by noting that ENIAC could "calculate the trajectory of a projectile in less than the time of flight of the projectile." In Tabor's words, this speed made possible "the solution of mathematical problems which have not heretofore been attacked because of the enormous amount of calculation required to obtain a solution."

In October 1981, almost an entire issue of the *Annals of the History of Computing* was devoted to an analysis of ENIAC by Arthur and Alice Burks and to commentaries by other participants in that machine's invention and by other innovators and historians. The Burkses concluded that Atanasoff's machine was "the first electronic computer," but that it was a special-purpose machine. ENIAC, they assert, was "the first general-purpose electronic computer."

If ENIAC was a computer even though it did not embody the concept of the stored-program, then there are grounds for considering the ASCC/Mark I to have been a computer too. Both machines were based on the concept of a controlled or directed sequence of arithmetical operations. However, the sequence of operations for Mark I was written as a external program on punched tape. ENIAC was programmed internally, by linking its various parts electrically by means of cables attached to plugboards.

To judge from Lewis Tabor's description, programming the ASCC/Mark I was very much like writing a program for a later computer in machine language. In this sense at least, Mark I may have been more like a latter-day computer than ENIAC was. Tabor stressed that setting up a complex problem was "a time consuming operation" requiring "careful planning of the programming, followed by a period of physically plugging patch-cords from buses to panels" and the subsequent chore of checking that the proper connections had been made. In contrast with Mark I, ENIAC was best suited, according to Tabor, for problems requiring "many solutions of an equation with changing parameters and initial conditions," particularly problems in which "a large number of calculations must be done to obtain each result."

Eckert, Mauchly, and their associates were fully aware that ENIAC's limitations arose from the fact that it had been designed primarily to perform ballistic calculations. Furthermore, as is rather well known, ENIAC was readily adapted to serve as a general-purpose computer. As Eckert reported in one of his lectures at the Moore School in July 1946, ENIAC's successor, EDVAC, was planned to be much more flexible, with "no cords, no plugs, and few switches." EDVAC was planned "to use the memory to hold the information electronically, and to feed those pieces of information which relate to programming from the memory into the control circuits," so that the machine would be sequenced in order "to perform its various operations." In presenting the new method of programming, Eckert used the programming of Mark I as his standard, showing in detail how the proposed machine would introduce real improvements over the programming of Mark I and would even remedy some of the earlier machine's defects. His remarks reinforce the conclusion that Mark I was *like* a computer.

There can be no simple answer to the question "What was the first 'computer'?" or to the related question "Who invented the 'computer'?" Each historian and computer scientist must decide for himself or herself whether any or all of the early "automatic," "sequenced," and program "controlled" machines are to be considered computers or precursors thereof. Certainly both Mark I and ENIAC had features that one associates with today's computers.

Most historians of the computer agree that the computer as we know it today was produced through the efforts of many heroic individuals and with the use of many technologies. In any case, Howard Aiken merits special credit for conceiving a machine that "would execute an arbitrary sequence of operations specified by a program" and then convincing IBM to design and construct a machine that would perform the operations in the way that he specified. It is a fact of history that Aiken's Mark I opened the eyes of many to the possibilities of large-scale programmed automatic computing. Actual witnesses to the developments of the mid 1940s, as we have seen, agree that its dedication inaugurated the computer age. Aiken's importance in the history of the computer need not be burdened with questions of whether or not his machines were "really" computers and, accordingly, whether or not in some sense he "invented" *the* "computer."

# Appendix I
# The Harvard Computation Laboratory during the 1950s

Some of the major activities of the Computation Lab during the early 1950s are recorded in extensive progress reports published for limited distribution. These reports record the development of Mark IV and its applications, giving details of the various phases of design, construction, testing, and operation of that machine and its components.

More than 150 of the approximately 500 pages of "Progress Report No. 18 (Covering Period: 10 May 1951–10 Aug. 1951)," titled "Investigation for Design of Digital Calculating Machinery (Contract: AF 33(038)-9461)," are devoted to circuit diagrams. Details are given concerning the divider, the magnetic drum unit, the tape read-record system, and the sequence unit.

"Progress Report 19 (Covering Period: 10 August 1951–10 Nov. 1951)," titled "Investigations for Design with Detail Calculating Machinery," contains further detailed descriptions of the testing, design, and construction (including changes in magnetic design) of the Mark IV coding machine, tests of the accumulator, and so on.

Progress Report No. 20 (Covering Period: 1 March 1951–1 November 1951), titled "Machine Solution of Problems for the Office of Air Research," contains reports on seven "AF" problems, including "Coordinate Determination of Ground Points on Aerial Photographs" (AF Problem No. 48, Problem Report No. 25), "Computation of the Intensities of Vibrational Spectra in Electronic Bands in Diatomic Molecules" (AF Problem No. 56, Problem Report No. 27), "Solution of a Reciprocal Equation of the Eighteenth Degree with Real Coefficients" (AF Problem No. 36, Problem Report No. 28), and "Reduction of Data Obtained from the Tracking of an Airplane" (AF Problem No. 61, Problem Report No. 29). All these problems were solved on Mark I, the only working computer in the laboratory. Each report gives both methods and solutions. For example, "Progress Report No. 18 (10 May

1951–10 August 1951)," titled "Investigations for Design of Digital Machinery," deals with Mark IV's divider, drum unit, tape read-record system, sequence unit, and codes. In this installment, references are made to practical problems for the Air Force—e.g., how to determine the coordinates of ground points from a series of overlapping photographs made by an airplane "flying at varying heights, where the input data consists of the height of the airplane for each photograph, the longitude and latitude of the airplane, and the tilt, swing, azimuth of the photographs" and how to "reduce and correlate data obtained from the tracking of an airplane."

Present-day readers may be puzzled by the Air Force's sponsorship of Charles Mannebach's method "for computing the Schrödinger matrices corresponding to vibrational transitions occurring along with given electronic transitions in certain diatomic molecules." The report introduces this topic with a statement that the matrix may be of general interest since "by squaring its elements a new matrix is generated, each row of which gives the relative probabilities of emission transitions to various final vibrational states for a given initial vibrational state." In no way was this problem related to practical concerns of the Air Force, even potentially. Howard Aiken's willingness to solve this problem is an index of his continuing concern for the computing needs of the scientific community. Furthermore, in those days the US armed forces sponsored a considerable amount of pure science or basic research, both in the United States and elsewhere.

Progress Report No. 20 also described and gave the status of each of seventeen "problems on Mark I as of 1 November 1951." It observed that during the preceding year "extensive changes" have been introduced, resulting "in the rebuilding of a large proportion of the machine." (The "new multiplier" had not as yet been completed.) These seventeen problems were undertaken for "Bell Aircraft Company, Sylvania Company, Aircraft Laboratory WADC, Engineer Research and Development Laboratory WADS, Pennsylvania State College, Syracuse University, General Precision Laboratory, Dr. A. Walther (Germany), Flight Research Laboratory WADC, Professor C. Mannebach (Belgium), Cambridge Research Laboratories, Aeromedical Laboratory WADC, Eglin Field, and a certain number assigned to Harvard University." The latter included "tables of Error Function" (published as volume 23 of the Annals of the Computation Laboratory), "Conduction of Heat" (described as a "low priority base load problem"), and "Behavior of Water Particles under Wet-Compression in Axial Compressors."

A similar list of problems appears in "Progress Report No. 24 (Covering Period: 10 August 1952–10 November 1952)," but there is a major difference. Since Mark IV was now undergoing testing by actual use before delivery to the Air Force, this status report of problems (prepared by John Harr) was titled "Status of Problems on Mark I and Mark IV as of 1 November 1952." Problem No. 17, for the Bell Aircraft Corporation, had now been transferred to Mark IV, and "several runs [had] been made." The new problems included one for "Dr. G. Swain (M.I.T.)" (said to require "the solution of approximately eighty non-linear simultaneous equations" and to be "in progress on Mark IV"), "a Perkin Elmer Co. problem" calling for "tracing of skew rays through an optical lens design," and a Harvard project intended to produce tables of inverse circular and hyperbolic functions.

A picture of the complex network of research, development, teaching, training, operation in problem solving, and entrepreneurship of the Computer Laboratory emerges from these reports. Aiken had, in effect, realized his dream of a private and quasi-independent training center and computing service within the structure of Harvard University. During the decade and a half when Aiken ran the Comp Lab, its output was prodigious—almost unbelievably so.

The Comp Lab's best-known publications are the 40 volumes of the Annals of the Computation Laboratory of Harvard University, mostly fat volumes of tables but also including the manuals of operation for Mark I, Mark II, and Mark III and proceedings of several symposia. There were also 60 volumes of Progress Reports to the Air Force (AF-1 through AF-60), produced between May 1948 and December 1960. In addition, from April 1949 to July 1959 Aiken's group produced 123 Problem Reports of the Computation Laboratory as supplements to the Progress Reports to the Air Force. Furthermore, between September 1952 and Aiken's retirement in 1961, 29 volumes of Reports on the Theory of Switching (BL-1 through BL-29) were completed for the Bell Telephone Laboratories. After Aiken's retirement, volumes BL-30 through BL-41 were produced under the direction of Anthony Oettinger and Hao Wang. And seven volumes of Progress Reports on Automatic Data Processing were produced between 1955 and 1958 for the Electronic Steering Committee of the American Gas Association and the Edison Electric Institute.

While directing the research for some 140 publications,[1] Aiken taught full-time and supervised 18 doctoral dissertations. In addition, he was responsible for the design and construction of three large computing machines, pioneering the use of magnetic-drum storage and magnetic core memories. He also hosted a constant stream of visitors, many from abroad, and gave frequent public lectures and speeches. The sheer magnitude of this activity leaves one breathless.

1. Gerard Salton notes that this number does not include the problem reports for the Air Force, most of which are quite short. He calls particular attention to Progress Report AF-23 (1952), which contains "what is likely to have been the first programming text, covering Mark IV coding." Salton also points out that the reports for the American Gas Institute and Edison Electric Institute demonstrate "Aiken's interest in non-numeric computer applications far in advance of anyone else." See Gerard Salton, "Aiken's Children," *Abacus* 1 (1984), no. 3: 28–34.

# *Sources*

## *Archival Sources*

This volume is based primarily on correspondence, memoranda, transcripts of interviews, and other documents (primarily in the Harvard University Archives and in the IBM Archives). A second major source of information is a series of interviews (and personal communications, including telephone conversations) conducted by the author. Copies of the primary documents and tapes of interviews and conversations have been deposited in the Harvard University Archives. Duplicate copies of some of the major sources have been deposited in the Charles Babbage Institute at the University of Minnesota. The following abbreviations are used for the archives.

CBIa: Charles Babbage Institute archives
HUa: Harvard University Archives
IBMa: archives of the International Business Machines Corporation.

Some of the Aiken material in HUa is in the Aiken files, some in the files of the Computation Laboratory. Other important Aiken documents are in an archive transferred to HUa by Anthony Oettinger in June 1997 and in the HUa's collection of the papers of Harlow Shapley. Additional manuscript sources in HUa are the President's Papers and the Papers of the School of Engineering.

The Aiken papers in HUa (especially those in the Oettinger archive) include drafts and transcripts of speeches and lectures and partially edited texts of articles that seem ready for publication. Excerpts from some of these are published in the companion volume, *Makin' Numbers: Howard Aiken and the Computer* (MIT Press, 1999).

Two archival sources of great importance are Gregory W. Welch's undergraduate honors thesis (Harvard University, 1986) Computer Scientist Howard Hathaway Aiken—Reactionary or Visionary? (on deposit in the Department of History of Science at Harvard University and in HUa) and the records of the celebrations of Aiken and his four machines at Pioneer Day, a feature of the National Computer Conference of the American Federation of Information Processing Societies held at Anaheim in May 1983. The meetings were organized by Robert Campbell, Richard Bloch, and I. Bernard Cohen. Tapes and transcripts are on deposit in CBIA.

## Publications by Aiken

Howard H. Aiken, "Trilinear Coordinates," *Journal of Applied Physics* 8 (1937) 470–472.[1]

[Howard H. Aiken and Grace Hopper], *A Manual of Operation for the Automatic Sequence Controlled Calculator.* Annals of the Computation Laboratory of Harvard University, volume 1, 1946. Reprint: MIT Press, 1985 (Charles Babbage Institute Reprint Series for the History of Computing, volume 8).

Howard H. Aiken and Grace Hopper, "The Automatic Sequence Controlled Calculator," *Electrical Engineering* 65 (1946): 384–391, 449–554.

Howard H. Aiken, "The Automatic Sequence Controlled Calculator" (Lecture 13, given 16 July 1946) and "Electro-Mechanical Tables of the Elementary Functions" (Lecture 14, given 17 July 1946), Moore School of Electrical Engineering, University of Pennsylvania. Reprinted in *The Moore School Lectures*, ed. M. Campbell-Kelly and M. Williams (MIT Press, 1985).

Howard H. Aiken, "Proposed Automatic Calculating Machine (previously unpublished memorandum)," *IEEE Spectrum* (August 1964): 62–69.[2]

## Annals of the Computation Laboratory of Harvard University

A rich record of the activities of the Computation Laboratory is provided by the 40 volumes of the Annals of the Computation Laboratory of Harvard University, published under Aiken's direction and editorship.

## Progress Reports and Related Items[3]

Progress Reports to the Air Force (AF-1–AF-60, May 1948–December 1960)

Problem Reports of the Computation Laboratory (123 reports, April 1949–July 1959)

---

1. This appears to have been Aiken's first published work. So far as I have been able to determine, this is the only work that Aiken composed on his own for publication.
2. For a more exact presentation of Aiken's memorandum, see *Makin' Numbers*.
3. For details, see appendix I.

Reports on the Theory of Switching (BL-1–BL-29) produced for Bell Telephone Laboratories (September 1952 until Aiken's retirement in 1961), continued (BL-30 to BL-41) under the direction of Anthony Oettinger and Hao Wang; and the 7 volumes of "Progress Reports on Automatic Data Processing" (1955–1958), produced for the Electronic Steering Committee of the American Gas Association and the Edison Electric Institute.

## Forewords and Afterwords

Valuable information may be found in the forewords (or prefaces) and afterwords (or conclusions) to the four symposium volumes Aiken edited and in the proceedings of the UNESCO conference in June 1959:

Proceedings of a Symposium on Large-Scale Digital Calculating Machinery (Annals of the Computation Laboratory of Harvard University, volume 16, 1948). Reprint: MIT Press, 1985 (Charles Babbage Institute Reprint Series for the History of Computing, volume 7).

Proceedings of a Second Symposium on Large-Scale Digital Calculating Machinery (Annals of the Computation Laboratory of Harvard University, volume 26, 1951).

Proceedings of an International Symposium on the Theory of Switching (Annals of the Computation Laboratory of Harvard University, volumes 29 and 30, 1959).

Aiken, Howard H., and William F. Main, eds. *Switching Theory in Space Technology*. Stanford University Press, 1963.

Information Processing 1959: Proceedings of the International Conference on Information Processing. UNESCO, 1959; Oldenbourg, 1960; Butterworth, 1960.

## Other Sources

Interview conducted with Aiken by Henry Tropp and I. Bernard Cohen at Aiken's home, Fort Lauderdale, Florida, 23 and 24 February 1973. Tapes and a transcript are on deposit in HUa, and a copy of the transcript is in CBIa.

Memorial Ceremony for Aiken, Memorial Church, Harvard University, 7 May 1973. Speakers: Harvey Brooks, Kenneth Iverson, Robert Ashenhurst, Anthony Oettinger, Jacqueline Sill, Col. George R. Weinbrenner. Copies of the talks by Iverson, Weinbrenner, Ashenhurst, and Sill are on deposit in the Aiken archives in HUa, where there is also a printed version for private distribution.

Interview with Francis Hamilton, conducted by Larry M. Saphire on 22 August 1967 as part of the IBM Oral History of Computer Technology. A transcript is available in IBMa, and copies have been deposited in HUa and CBIa.

Hamilton memo: a 15-page, double-spaced, typewritten, historical memorandum, dated 21 August 1944, without any title or heading, concerning the design and construction of the IBM ASCC/Harvard Mark I. This document, written shortly after the dedication, is available in IBMa; copies have been deposited in HUa and CBIa.

## Published Sources

Ashenhurst, Robert L. "Howard Hathaway Aiken, 1900–1973." *Communications of the Association for Computing Machinery* 16 (1973): 274.

Aspray, William. *John von Neumann and the Origins of Modern Computing.* MIT Press, 1990.

Aspray, William, ed. *Computing before Computers.* Iowa State University Press, 1990.

Aspray, William, and A. Burks, eds. *Papers of John von Neumann on Computing and Computer Theory.* MIT Press, 1987.

Babbage, Henry Prevost, ed. *Babbage's Calculating Engines: Being a Collection of Papers Relating to Them, Their History, and Construction* (London: E. and F. N. Spon, 1889; MIT Press, 1982).

Bashe, Charles, Lyle R. Johnson, John H. Palmer, and Emerson W. Pugh. *IBM's Early Computers.* MIT Press, 1985.

Belden, T. G., and M. R. Belden. *The Lengthening Shadow.* Little, Brown, 1962.

Berkeley, Edmund C. *Giant Brains, or Machines That Think.* Wiley, 1949; Science Editions, 1961. (Berkeley, a former staff member of the Harvard Computation Laboratory, could write of Mark I from firsthand experience.)

Bernstein, Jeremy. *The Analytical Engine: Computers, Past, Present, and Future.* Random House, 1964.

Ceruzzi, Paul E. The Prehistory of the Digital Computer, 1935–1945: A Cross-Cultural Study. Doctoral dissertation, University of Kansas, 1981 (available from University Microfilms, Ann Arbor).

Ceruzzi, Paul E. *Reckoners: The Prehistory of the Digital Computer.* Greenwood, 1983.

Cohen, I. Bernard. "The Computer: A Case Study of Support by Government, especially the Military, of a New Science and Technology." *Science, Technology and the Military,* ed. E. Mendelsohn et al. Kluwer, 1988.

Cohen, I. Bernard. "Babbage and Aiken." *Annals of the History of Computing* 10 (1988): 171–193.

Cohen, I. Bernard. "Howard H. Aiken and the Computer." In *A History of Scientific Computing*, ed. S. Nash. ACM Press and Addison-Wesley, 1990.

Cohen, I. Bernard. "Howard H. Aiken, Harvard University, and IBM: Cooperation and Conflict." In *Science at Harvard University: Historical Perspectives*, ed. C. Elliott and M. Rossiter. Lehigh University Press, 1992.

Cohen, I. Bernard. "Howard Aiken and the Beginnings of Computer Science." *CWI Quarterly* (Centrum voor Wiskunde en Informatica, Amsterdam) 4 (1990): 303–324.

Cortada, James. *Historical Dictionary of Data Processing*, three volumes (Greenwood, 1987). See especially "Harvard Mark I, II, III, IV" in the Technology volume and "Aiken, Howard Hathaway" in the Biographies volume.

Getting, Ivan. *All in a Lifetime: Science in the Defense of Democracy*. Vantage, 1989.

Goldstine, Herman. *The Computer from Pascal to von Neumann*. Princeton University Press, 1972.

Grosch, Herbert R. J. *Computer: Bit Slices from a Life*. Third Millennium Books, 1991.

Metropolis, N., J. Howlett, and Gian-Carlo Rota, eds. *A History of Computing in the Twentieth Century*. Academic Press, 1980.

Oettinger, Anthony G. "Retiring Computer Pioneer–Howard Aiken." *Communications of the ACM* 5 (1962): 298–299.

Oettinger, Anthony G., I. Bernard Cohen, Wassily W. Leontief, and Harry R. Mimno. "Howard Hathaway Aiken." *The Gazette* (Harvard University) 69, no. 37 (7 June 1974): 8.

Office of Charles and Ray Eames. *A Computer Perspective*, second edition. Harvard University Press, 1990.

Pugh, Emerson. *Building IBM: Shaping an Industry and Its Technology*. MIT Press, 1995.

Ralston, Anthony, and Edwin Reilly Jr., eds. *Encyclopedia of Computer Science and Engineering*, second edition. Van Nostrand Reinhold, 1983.

Randell, Brian, ed. *The Origins of Digital Computers: Selected Papers*, third edition. Springer-Verlag, 1975.

Stern, Nancy. *From ENIAC to UNIVAC: An Appraisal of the Eckert-Mauchly Computers*. Digital Press, 1981.

Tropp, Henry. "Aiken, Howard." In *Encyclopedia of Computer Science and Engineering*, second edition, ed. A. Ralston and E. Reilly Jr. Van Nostrand Reinhold, 1983.

Watson, Thomas J., Jr. *Father and Son & Co.: My Life at IBM and Beyond* (Bantam, 1990).

Wilkes, Maurice V. *Automatic Digital Calculators*. Methuen, 1956.

Wilkes, Maurice V. *Memoirs of a Computer Pioneer*. MIT Press, 1985.

Williams, Michael R. *A History of Computing Technology*. Prentice-Hall, 1985. Revised edition: ACM Press, 1997.

Zuse, Konrad. *The Computer—My Life*. Springer-Verlag, 1991.

## Source Notes for Individual Chapters

### Chapter 1

The general information about Aiken presented in this chapter is based to some degree on the author's personal knowledge.

The anecdote about Aiken and Bundy is told by Garrett Birkhoff in Metropolis et al., *A History of Computing in the Twentieth Century*.

The quotation from Brooks is from a talk given at the National Computer Conference of the American Federation of Information Processing Societies at Anaheim in May 1983. A tape and a transcript are available in HUa and a transcript in CBIA; a modified version appears in *Makin' Numbers*.

Zuse's account of the high opinion of his colleagues concerning Aiken's calculator is on page 127 of his book *The Computer—My Life*.

The story of Wilkes's two encounters with Aiken comes from Wilkes's 1983 Pioneer Day talk, supplemented by personal communications. A modified version of the talk appears in *Makin' Numbers*. See also Wilkes, *Memoirs of a Computer Pioneer*.

The presentation of Auerbach's recollections about Aiken is based on discussions between the author and Auerbach, supplemented by transcripts of two interviews (on deposit in CBIa), and on an obituary of Auerbach by Erich Weiss (*Annals of the History of Computing* 15 (1993): 60–61).

Harr's recollections of his interviews with Aiken come from his 1983 Pioneer Day talk.

The statement by Paul Ceruzzi is from his doctoral dissertation, Prehistory of the Computer.

### Chapter 2

This chapter is based on the 1973 Tropp-Cohen interview with Aiken, supplemented by information given to me by Mary Aiken and by documents in the Harvard University Archives.

All the letters to and from Aiken quoted here are in the Aiken files in HUa.

The information on Edward Bennett is from volume 3 of *Who Was Who in America*. Aiken's account of Bennett's philosophy of technological development is from a talk given by Aiken, once in Sweden and again in Germany, in 1959; complete texts are available in HUa (Oettinger archive); portions are included in *Makin' Numbers*.

Aiken's autobiographical letter to Weaver is in the Aiken files in HUa; a copy is in CBIa.

The information about Bert Little, here and elsewhere in this volume, is derived from my personal reminiscences, from conversations with Mrs. Little, with Betty Jennings (Cohen), and from Little's published Harvard class reports.

The source of my knowledge concerning Aiken's noonday lunches with his mother is a 1996 conversation with Betty Jennings (Cohen).

A copy of Aiken's curriculum vitae is among the Aiken papers in HUa. Major events in Aiken's life and career are listed in successive editions of *Who's Who in America* and in the cumulative volumes of *Who Was Who,* and also in the 1947 edition of *Current Biography* (pp. 5–7). The dates given by Aiken in these sources do not always agree with one another.

## Chapter 3

The information about Aiken's early years as a student in the Harvard Physics Department is from conversations with Ronald W. P. King and Edward Mills Purcell.

The information about the structure of the Harvard Physics Department is from various Harvard publications, from the university's annual reports, and from the author's recollections.

Aiken's doctoral dissertation, Theory of Space Charge Conduction, is on deposit in HUa. The Oettinger archive (in the Aiken files in HUa) contains an admirable four-page summary by Oettinger of the contents of this dissertation; a copy of this summary is on deposit in CBIa.

Aiken's early teaching career is recorded in Harvard's course catalogues and annual reports. In those days at Harvard, the course catalogue for each year was bound into a volume with reports of various departments and other materials. The number of students in each course was reported.

Hooper's recollections are taken from personal correspondence and from transcripts of conversations and telephone calls, some of which are on deposit in HUa.

Stone's account of the support of Aiken's candidacy by the Harvard Mathematics Department is taken from his article "Men and Institutions," *Graduate Studies, Texas Technical College* 13 (1976): 19.

On Kennelly, see volume 7 of the *Dictionary of Scientific Biography.*

Mimno's recollections of Aiken's lobbying activities are from a personal memorandum he wrote at Anthony Oettinger's request when Oettinger was preparing a "minute" on the life and services of Aiken for the records of the Harvard Faculty of Arts and Sciences. A copy is available in HUa.

## Chapter 4

On pre-computer methods of calculation, see *Computing before Computers* (ed. Aspray); Williams, *History of Computing Technology;* and the early chapters of Bashe et al., *IBM's Early Computers.*

On Eckert and the Watson Laboratory, see Jean Ford Brennan, *The IBM Watson Laboratory at Columbia University: A History* (IBM Corporation, 1971). A transcript of a lengthy interview with Eckert is in IBMa; copies are on file in HUa and CBIa.

On Comrie, see volume 3 of the *Dictionary of Scientific Biography;* see also *Obituary Notices of Fellows of the Royal Society* 8 (1952): 97–107.

On Aiken and the relays, see Getting, *All in a Lifetime.*

Bernstein describes his encounter with Aiken in *The Analytical Engine.*

## Chapter 5

The story of Aiken's encounter with Chase is based on the 1973 Tropp-Cohen interview with Aiken and on Chase's recollections in "History of Mechanical Computing Machinery," *Annals of the History of Computing* 2 (1980): 198–226. The latter is an adaptation (with a foreword by I. B. Cohen) of Chase's 1952 presentation in *Proceedings of the Association for Computing Machinery.* The foreword contains a brief biography of Chase. Aiken's later correspondence with Chase is in the Aiken files in HUa.

## Chapter 6

Aiken recalled Chase's suggestion in the 1973 interview.

The information on Brown is from an interview by Owen Gingerich and I. B. Cohen, supplemented by *Who's Who in America,* by *Who Was Who,* and by the annual catalogues of Harvard University.

Hamilton's historical memorandum is in IBMa; copies are in HUa and CBIa. On Shapley see *Who's Who in America* and *Who Was Who.* For a history of the increasing support of Aiken's project, as seen from an IBM point of view, see chapter 4 of *Makin' Numbers.*

On Watson, see the authorized biography by Belden and Belden, *The Lengthening Shadow;* see also chapter 2 of Pugh, *Building IBM.* On Bryce, see *Building IBM;* see also "The Light He Leaves Behind," *Think,* April 1949.

## Chapter 7

Several copies of Aiken's proposal are in HUa (among the Aiken papers, the president's files, and the files of the dean of engineering). An edited version was published in the August 1964 *IEEE Spectrum* (pp. 62–69) and later reprinted in Randell, *Origins of Digital Computers* (and elsewhere). An unedited and unaltered version is published in *Makin' Numbers.*

The discussion of "architecture" is based on the entry in the *Encyclopedia of Computer Science and Engineering* and on the presentation in Bashe et al., *IBM's Early Computers.* See also Gerrit Blaauw and Frederick P. Brooks Jr., *Computer Architecture, Concepts and Evolution* (IEEE Computer Society, 1997).

## Chapter 8

Most of this chapter is drawn from my article "Babbage and Aiken" in *Annals of the History of Computing.*

For Comrie's review, titled "Babbage's Dream, Come True," see *Nature* 158 (26 October 1946): 567–568. The articles by Aiken and Hopper (listed above) are reproduced in Randell, *Origins of Digital Computers.*

Sources    317

The quotations attributed to Bernstein are from his book *The Analytical Engine.*

The proceedings of the symposium at which Richard Babbage spoke (the 1947 Symposium on Large-Scale Digital Calculating Machinery) were published as volume 16 of the Annals of the Computation Laboratory of Harvard University.

Randell's comments appear on page 187 of *Origins of Digital Computers;* Hopper's recollections of Aiken and Babbage are quoted by Randell on page 422 of the same book.

The two sources of information on Babbage that Aiken cited in his proposal were D. Baxendall, "Calculating Machines and Instruments," in *Catalogue of the Collections in the Science Museum* (HMSO, 1926), and *Napier Tercentary Celebration,* ed. E. Horsburgh (Royal Society of Edinburgh, 1914; MIT Press, 1982).

## Chapter 9

Aiken's correspondence with Bryce and Hamilton's historical memorandum are in IBMa; copies are on deposit in HUa and in CBIa.

On IBM and the construction of the ASCC/Mark I see Charles Bashe's chapter in *Makin' Numbers,* a somewhat longer version of which is in HUa and CBa.

On the Columbia computing facility and its various names, see Jean Ford Brennan, *The IBM Watson Laboratory at Columbia University* (IBM Corporation, 1971).

The correspondence between Pym and Bryce is in IBMa; photocopies of major letters are in HUa and in CBIa.

McDowell's report to Bryce is in IBMa; a copy is on deposit in HUa.

Bryce's letter or memorandum to Hamilton is in IBMa; a copy is on deposit in HUa.

Copies of the preliminary and final agreements between Harvard and IBM are in IBMa and in HUa.

Westergaard's and Bryce's letters are in IBMa; copies are in HUa.

Aiken's activities in the summer of 1939 are described in the transcript of the Hamilton interview.

Aiken's letter of November 1940 is in IBMa; a copy is in HUa.

The information on the early machines and on the IBM method of partial products is from Bashe et al., *IBM's Early Computers.*

Durfee was interviewed by Larry Saphire on 22 August 1967 as part of the IBM Oral History of Computer Technology. A transcript is in IBMa; copies are on deposit in HUa and CBIa.

## Chapter 10

The discussions of methods of multiplication and of the operation of the IBM 601 are based on information provided to the author by Lyle Johnson and on Johnson's presentation of this subject in Bashe et al., *IBM's Early Computers.*

The latter volume is also the source of the statement concerning IBM machines' inability to perform division.

Bloch's discussion of multiplication is from his presentation at Aiken's 1947 symposium.

A transcript of the taped interview with Daly can be found in IBMa; a copy is on deposit in HUa.

## Chapter 11

The account of Aiken at IBM in the summer of 1939 is based on Hamilton's historical memorandum. The correspondence with Lake is in IBMa and HUa.

Durfee was interviewed by Larry Saphire on 22 August 1967 as part of the IBM Oral History of Computer Technology. A transcript is in IBMa; copies are on deposit in HUa and CBIa.

The information on Aiken and Campbell was provided by Campbell.

Why the new machine took so long to construct is discussed in Hamilton's historical memorandum and in Bashe's chapter in *Makin' Numbers*.

Hamilton expressed his view of the odd, problematic character of the new machine in the course of his 1967 interview with Larry Saphire.

Information on Campbell's input in the last phases of construction is to be found in Hamilton's historical memorandum.

The Shapley-Chase correspondence is in HUa.

The Conant-Watson correspondence is in IBMa and HUa.

## Chapter 12

The Conant-Watson correspondence is in IBMa and HUa. Watson's letter to Conant and Bryce's letter to Pierce are in IBMa; copies are in HUa. Watson's efforts to obtain the services of someone in electronics are detailed in Pugh, *Building IBM*.

The information on the first days of the new machine at Harvard is from conversations with Bob Campbell and with Dave Wheatland.

The information on the first programs and the early operation of the machine is from the logbook kept by Campbell and from discussions with him. See his chapter on Mark I in *Makin' Numbers*.

Photocopies of the early pages of the logbook are in HUa. The logbook itself is now in the Museum of American History of the Smithsonian Institution.

## Chapter 13

Most of this chapter is based on conversations with John Ladd. Tapes and transcripts of some of these are in HUa. The letters from Neisser and Monahan and carbons of Aiken's letters to Boyd and to Giletti are in the Aiken files in HUa. I have not found any information on these correspondents.

The information on Wright is from John Ladd and from a telephone conversation with Wright.

Shapley's correspondence with Aiken is in the Shapley files of HUa. The identification of the two officers responsible for assigning Aiken to Harvard and for the Navy's taking over (and staffing) the operation of Mark I is from Aiken's talk at the dedication; he repeated it in other talks.

## Chapter 14
The official record of the dedication of the SSEC was published in IBM's *Business Machines* (15 March 1948).

On IBM's wartime activities see Bashe et al., *IBM's Early Computers*.

Hamilton's statement about his war work is from the historical memorandum and from the 1967 interview.

The discussion of Campbell's contributions to the machine is based on personal communications and on Hamilton's 1944 memorandum. See also Campbell's chapter in *Makin' Numbers*.

Information on Walter Lemmon is available in the volumes of *Who's Who in America* and *Who Was Who*. There is a large mass of correspondence between Lemmon and Shapley in the Shapley files in HUa.

## Chapter 15
Westergaard's correspondence with Watson is in IBMa and HUa. Westergaard's comments on his letter are in the President's Papers in HUa. Information on the proposed symposium can be found in letters exchanged by Westergaard, Conant, and Shapley. Watson's reply to Westergaard is in IBMa and in HUa.

The information on Shapley's activities after Westergaard's letter is based on correspondence in the Shapley files in HUa. The Shapley-Brown correspondence is in the Shapley papers in HUa.

Copies of Aiken's letters to Bryce and Lake are in HUa. Watson's letter to Aiken is in HUa.

Several copies of the IBM brochure are in HUa and in CBIa.

Brooks's comments are quoted from his 1983 Pioneer Day talk, partly presented in his chapter of *Makin' Numbers*. *IBM's Early Computers* documents L. Johnson's use of the term "system architecture."

The student's handbook was made available to me by Jack Palmer (an M.S. student in the early years of Aiken's Harvard program in computer science), who kindly consented to depositing his copy in HUa; a photocopy is in CBIa. The cover page reads: "STUDENTS' HANDBOOK for the AUTOMATIC SEQUENCE CONTROLLED CALCULATOR, MARK I. For Students of Applied Science 211 and 212. Computation Laboratory, Harvard University, September, 1949."

## Chapter 16
The details concerning the size and number of parts are from the IBM brochure, a copy of which is in HUa. Technical details of all aspects of Mark I are given in the Manual of Operation. On the architecture and the programming of Mark I, see the chapters by Campbell and by Bloch in *Makin' Numbers*.

Campbell describes the difficulties he encountered in using the new calculator after it had been installed at Harvard in *Makin' Numbers*. The presentation here draws on that chapter and on conversations with Campbell.

## Chapter 17

The information on Palmer's experience programming Mark I is from a series of conversations I had with him.

The discussion of a name for the new machine is based on conversations with Campbell.

The information on Baker's problems is from conversations and interviews with him.

Information on the Navy staffing of Mark I is based on the Manual of Operation, on the chapters by Campbell and Bloch in *Makin' Numbers,* and on conversations with Campbell.

Some of the information on Seeber is from an interview conducted by Larry Saphire, from his published class reports (Harvard College), and from Campbell's recollections.

On the programming of the implosion problem, see Bloch's chapter in *Makin' Numbers*. The correspondence relative to this event is in HUa.

The Metropolis-Nelson report appears in *A History of Computing in the Twentieth Century,* ed. Metropolis et al.

The HUa contain a partly edited transcript of the talk "Computers—Past and Future," given by Aiken at an Executive Seminar on the Use of High-Speed Calculators for the Solution of Naval Problems (7–9 September 1960, David Taylor Model Basin, Washington).

## Chapter 18

On Grosch's experience in computation, see his book *Computer: Bit Slices from a Life*.

Campbell's description of the test problems he devised for Mark I appears in his chapter on Mark I in *Makin' Numbers*.

Little's recollections of Aiken and the number 23 are from a document in the Aiken files in HUa.

## Chapter 19

The information on Salzberg is from discussions with Bloch. See also Bloch's chapter in *Makin' Numbers* and various autobiographical memoirs deposited in HUa.

The information on the tables produced by Comrie for the British Association is from the introductory portions of the published tables.

## Chapter 20

The information about the Columbia program is from a series of pamphlets issued by Columbia University describing the courses of instruction and the facilities available. Copies have been deposited in CBIa and HUa.

Aiken's pessimism about the availability of mathematicians is discussed in appendix F.

On the financial support of Aiken's program, see Gregory Welch's essay in *Makin' Numbers*.

The course offerings in the Harvard program are listed in the annual catalogues of courses.

Copies of the 1949 "Students' Handbook" are in HUa and in CBIa.

A copy of the final exam for Mimno's course (containing the spelling ENIAK) is in HUa.

On the courses and how they were taught, see the chapters by Brooks and by Calingaert in *Makin' Numbers*.

The reports of the various committees are in HUa.

## Chapter 21

This chapter draws heavily on two taped interviews with Cuthbert Hurd, one conducted by Henry Tropp and Robina Mapstone in October 1972 and one by I. B. Cohen in August 1983. Transcriptions of both are on deposit in HUa and in CBIa. The chapter also draws on a number of telephone conversations between the Cohen and Hurd in 1996 and 1997; tapes and transcripts of some of these are in HUa. On Hurd, see Eric Weiss, "Eloge," *Annals of the History of Computing* nn (1997): nnn–nnn.

The "history wall" can be seen in *A Computer Perspective*, by the Office of Charles and Ray Eames.

## Chapter 22

This chapter draws heavily on the recollections of a number of Aiken's former students and associates and from conversations with James Baker, Frederick Brooks, Richard Bloch, Robert Burns, Robert Campbell, Peter Calingaert, John Edsall, Grace Hopper, Wassily Leontief, Anthony Oettinger, John Palmer, and others. Among the former students who have written about their experiences at Harvard and in the Comp Lab are Brooks, Oettinger, Robert Ashenhurst, and Gerard Salton. For important recollections, see the transcripts of talks given at the Aiken Memorial Service by Ashenhurst, Kenneth Iverson, Col. George Weinbrenner, and Jacqueline Sill, and the 1983 Pioneer Day talks given by Bloch, Oettinger, John Harr, and Maurice Wilkes.

I have also made use of a draft essay by John Harr on Mark III and Mark IV, on deposit in HUa. See the chapters in *Makin' Numbers* by Welch, Brooks, Calingaert, Hopper, Oettinger, and Wilkes.

Galbraith cites Leontief's input-output economics in his 1987 book *Economics in Perspective*.

## Appendix B

Three different versions of Aiken's talk at the dedication are in HUa. There is an early draft in the form of a typescript with handwritten corrections. This leads to a second typed version, evidently intended to be used in the coming ceremonies. Both of these are in HUa in the Oettinger files. The third version, existing in several copies in HUa, is part of a transcript of all the talks given at the ceremonies, including—in addition to Aiken's talk—Conant's introduction, Westergaard's remarks, and Watson's closing speech. This transcript was based on a stenotype record made by a professional recording service—in those days, there were no wire or tape recorders. Here I present Aiken's talk from this transcript, correcting some obvious errors in transcription and adding some punctuation.

## *Appendix C*
This appendix reproduces the original document, which is in HUa.

## *Appendix D*
The quotations from Gill are from his article on the stored program in the *Encyclopedia of Computer Science and Engineering*.

## *Appendix E*
On Mark II, see Campbell's chapters in *Makin' Numbers* and in *Proceedings of the First Symposium of Large-Scale Digital Calculating Machinery* (Annals of the Computation Laboratory, volume 16). A manual of operation for Mark II, entitled Description of a Relay Calculator, was published as volume 24 of the Annals. For an important contemporaneous view of Mark II, see chapter 13 of Wilkes, *Automatic Digital Calculators*.

The correspondence between Conant and Watson relative to Mark II is in HUa.

On Mark III, see *Description of a Magnetic Drum Calculator* (Annals of the Computation Laboratory, volume 25). See also Benjamin Moore, "The Mark III Calculator," in *Proceedings of the Second Symposium on Large-Scale Digital Calculating Machinery* (Annals of the Computation Laboratory, volume 26).

A series of discussions with users of these two machines at Dahlgren were recorded on tape by J. A. N. Lee and Henry Tropp; a partially edited transcript is available in HUa and CBIa.

Elliott's report of a conversation with Aiken appears in *Report on High Speed Automatic Calculating-Machines, 22–25 June 1949* (issued by the University Mathematics Laboratory, Cambridge, with the cooperation of the Ministry of Supply, January 1950; reprinted in *The Early British Computer Conferences*, ed. M. Williams and M. Campbell-Kelly (MIT Press, 1989).

There was no officially published manual of operation for Mark IV, as there had been for the three earlier machines, but a privately circulated work, about 150 pages in length, was evidently prepared for users of the machine. Reproduced from a typewritten original (perhaps by some kind of mimeograph machine), this work bore the title Manual of Operation for the Harvard Magnetic Drum Calculator (Mark IV). The cover page presented the title, the name of the author (Norman B. Solomon), and the date (July 1957). There appears to be no copy of this work in the Harvard library system or in HUa.

The only copy I have encountered is in the private collection of Robert Campbell, who kindly brought it to my attention and who plans to donate his copy to HUa.

Plenty of technical information about Mark IV is available in the above-mentioned Progress Reports.

For Mark III and Mark IV, in addition to the manuals of operation, I have drawn on John Harr's 1983 Pioneer Day talk and on an essay based on that presentation, a copy of which is in HUa. The descriptive quotations concerning the features of Mark III and Mark IV are taken from Harr's essay and, in the case of Mark IV, from the manual of operation.

Aiken described some of the features of Mark IV in a talk given in 1952 at the American Management Association Office Management Conference; see *Makin' Numbers*.

## *Appendix F*

On the work of Subcommittee Z, see Stern, *From ENIAC to UNIVAC* and Aspray, *John von Neumann and the Origins of Modern Computing*. A copy of the draft report of the subcommittee is in HUa; a copy is in CBIa.

The text of Cannon's memorandum is in the Eleuthyrian Mills–Hagley Institute in Wilmington; I used a transcript made by Paul Ceruzzi, who very kindly made it available to me.

Auerbach's remarks are from a 1978 interview by Nancy Stern, a transcript of which is in CBIa. A copy of the portion dealing with Aiken is in HUa.

Von Neumann's talk, "The Future of High-Speed Computing," is reprinted in *Papers of John von Neumann on Computing and Computer Theory*, ed. Aspray and Burks.

## *Appendix H*

Wilkes's remarks on EDSAC are from his article on that machine in the *Encyclopedia of Computer Science and Engineering*. On EDVAC, see the article on it in the same encyclopedia. See also Stern, *From ENIAC to UNIVAC*, and *Papers of John von Neumann on Computing and Computer Theory*, ed. Aspray and Burks. On the Manchester machines, see Simon Lavington, *Early British Computers: The Story of Vintage Computers and the People Who Built Them* (Manchester University Press and Digital Press, 1980).

On Zuse's and Atanasoff's machines, see Ceruzzi, *Reckoners*, and Aspray, *Computing before Computers*.

For comments by Atanasoff on his own contributions, see "The Advent of Electronic Digital Computing," *Annals of the History of Computing* 6 (1984) 229–282; see also Williams, *History of Computer Technology*, which is especially good on Stibitz's machines.

Sale's claims for Colossus are reported and discussed by Christian Tyler in "Colossus Faces Rebirth into a World of Dispute," *Financial Times* (London), 27–28 July 1966, pp. 1–2. Wilkes's remarks about ENIAC as a computer are quoted in that same article.

The Burkses' article supplements one by Arthur Burks: "The ENIAC: The First General Purpose Electronic Computer," *Annals of the History of Computing* 3 (1982): 310–399.

## Appendix I

For more on the output of the Comp Lab under Aiken's direction, see Gerard Salton, "Aiken's Children," *Abacus* 1 (1984), no. 3: 28–34.

# *Index*

Accumulators, 96, 149
Accuracy, 94, 150–151, 169–176
Addition, over-and-under, 87–88
Aiken, Agnes Montgomery, 99, 164, 241
Aiken, Daniel, 9
Aiken, Elizabeth, 241
Aiken, Margaret Emily Mierisch, 9
Aiken, Mary, 9, 118–119, 227, 241–242
Aiken Industries, 231–232
Alonso, Ramón, 229
Analytical engine, 63–64, 70–72
Architecture, computer, 59, 144
Arnold, Hubert, 161
ASCC. *See* Mark I computer
Ashenhurst, Robert, 185, 215–216, 221
Astronomy, 169–172
Auerbach, Isaac, 3, 221–222, 245

Babbage, Charles, 61–72, 150
Babbage, Henry Prevost, 62–63, 67
Babbage, Richard, 65
Baker, James, 108, 112–114, 160–161, 169–170, 173–175
Baker, John, 105
Bargmann, Valentine, 164–165
Baxandall, D., 66, 69–70
Beckjord, Walter, 13–14
Belden, M. R., 143
Belden, T. G., 143
Bell Telephone Laboratories, 44
Bennett, Edward, 12

Berkeley, Edmund, 162
Bernstein, Jeremy, 36, 64
Bessel, Friedrich Wilhelm, 177
Bessel functions, 177–183
Birkhoff, Garrett, 193, 214, 223, 226
Birkhoff, George, 82
Bloch, Richard, 89, 152, 161–165, 174, 179, 216, 233–234, 238–239
Branching, 67–68, 152
Brendel, Ruth, 162, 180
Bridgman, P. W., 82
Brooks, Frederick P., Jr., 1, 143–144, 211–212, 216–218, 222, 227–228, 239, 241
Brown, Ernest William, 46
Brown, Theodore, 45–47, 50, 74–75, 79, 81–82, 134, 136–139, 173, 193
Bryce, James Wares, 46–47, 50–54, 73–75, 78, 81–88, 91–94, 98, 100, 110, 128–129, 139, 142
Buck, Paul, 142, 144, 224
Bundy, McGeorge, 1, 229
Burns, Robert, 240
Business, computers and, 211–213

Calculators, 33–34
Calingaert, Peter, 205, 229
Campbell, Robert, 16, 67–68, 99–102, 111–114, 152–153, 160–164, 169–172, 174, 179, 216
Campbell-Kelly, Martin, 62
Carlson, Walter, 235–236

Celestial mechanics, 169–172
Ceruzzi, Paul, 6
Chaffee, E. L., 21–23, 109, 134, 164, 225
Charles Babbage Institute, 62
Chase, George C., 39–42, 45
Chase, George H., 102–105, 133
Claflin, William, 105, 135–137
Classification, knowledge through, 12
Cohen, Martin, 229
College years, Aiken's, 11–14
Columbia University, computation center at, 34
Computation Laboratory, Harvard University, 6, 201–205, 215–226
Computer science programs, 6, 185–196
Computing-Tabulating-Recording Co., 48–49
Comrie, L. J., 34, 46, 61, 68, 157–158, 180–182
Conant, James Bryant, 28, 78–81, 104–110, 121–122, 125, 131, 135
Consulting work, Aiken's, 232–236
Coolidge, Charles, 230
Counters, 97, 147
Cross-footing, 89
Cruft group, 21–22
Cryptography, 233–236
Cunningham, Leland, 173–174

Daly, George, 90–91
Deduction, knowledge through, 12
Dickinson, Halsey, 46
Difference engine, 63–64, 69–72
Differential equations, 28, 33, 150
Division, 77–78, 90–94
Donaldson, Paul, 111
Douglas, Edward, 103–104, 133
Durfee, Benjamin, 73–74, 80–81, 92–101, 111, 142, 171–172

Eckert, J. Presper, 3
Eckert, Wallace, 34, 46
EDSAC computer, 204

Edsall, John, 208–209
EDVAC computer, 204
Electromechanical systems, 40–41, 43–44
Elias, Peter, 188
Encryption, 233–236
ENIAC computer, 204
Esch, Robin, 205

Ferrier, David, 119–120
Furry, Wendell, 179–180

Galbraith, John Kenneth, 206
Garrett, George, 232
Genesis (company), 233
Getting, Ivan, 35
Gingerich, Owen, 206
Goheen, Harry, 162
Graduate school, Aiken in, 18, 21–31
Grosch, Herbert, 34, 169–170

Haar, John, 4
Halbrook, Thomas, 14, 18
Hamilton, Francis, 46, 73–84, 91–98, 100–103, 142, 171
Hartlein, Albert, 193
Hawkins, Robert, 99–100, 111–112, 160, 162, 180, 240
Hayward, John, 133–134
Henderson, Derek, 229
High school, Aiken in, 10–11
Hollerith, Herman, 64
Honors and decorations, Aiken's, 220
Hooper, James, 25–28
Hopper, Grace, 152, 161, 163, 169, 180, 203, 241, 243
Horsburgh, E. M., 66, 69–71
Householder, Alton, 245
Humphrey, Jo Walker, 15, 18
Hunt, Theodore, 21
Huntington, E. V., 82
Hurd, Cuthbert, 123, 197–200, 227–228, 232–236
Hutchins, Robert Maynard, 17

IBM Corporation, 45–59, 63, 121–129, 197–200
Implosion, calculations for, 164–166
Indianapolis Light and Heat Co., 10
Information Security Corporation, 233
Instructions, 159–160
Iverson, Kenneth, 193, 205–208, 216–217, 237

Jakobson, Roman, 208
Jennings, Betty, 19, 206, 241
Johnson, Anthony, 10
Johnson, Lyle, 144

Kemble, E. C., 82
Kennelly, Arthur Edwin, 29–30, 222
Kennelly-Heaviside layer experiments, 30
King, Ronald, 21–22, 112, 160, 218
Knowlton, Ruth, 161
Kunz, Kaiser, 188

Ladd, John, 115–118
Lake, Clair, 49, 73, 76, 81–84, 95–101, 109, 139–142, 153–154
Lanza, Carmello, 66–67
Learson, T. Vincent, 198–199, 224
Lectures, Aiken's, 220–221
Lemmon, Walter, 125–128, 133–134, 136
Lenses, calculations for, 161
Leonard, Thomas, 11–12, 242
Leontief, Wassily, 158, 193, 205–208
Line Material Co., 13, 17
Little, Elbert (Bert), 18, 172, 187
Lochhart, Brooks, 162
Lockheed, 232
Loewner, Charles, 164–165

Madison Gas and Electric, 11, 13–14, 16
Magnetic-bubble technology, 233
Main, William, 232–233
Management style, Aiken's, 231–232

Mark I computer, 147–158
accuracy of, 94, 169–176
and Bessel functions, 177–183
coding and, 152, 159–167
construction of, 95–108
constants and, 77
decimal points and, 152
dedication of, 121–129
design of, 38
digits and 169–176
and errors, 153
IBM's support of, 45–51
idea for, 33–38
installation of, 109–114
Manual of Operation for, 64, 163
model of, 230
Monroe Calculating Machine proposal and, 39–42
output of, 149
panels of, 147–149
planning of, 73–86
precision of, 94, 169–176
programming of, 152, 159–167
proposal for, 55–58
reliability of, 153
separation of data and instructions in, 159
speed of, 156–158
technology considerations and, 43–44
Mark II computer, 162–163, 203–204
Mark III computer, 203
Mark IV computer, 203–204
Martin, Reino, 230
Mauchly, John, 3, 202
McCabe, Frank, 123–124
McDowell, W. Wallace, 75–76, 80–81, 83
McGrath, Richard, 231–232, 237
McKay, Gordon, 193
Methodology, knowledge through, 12
Metropolis, Nicholas, 166
Michelson, A. A., 17

Middleton, David, 188
Millikan, Robert Andrews, 17
Mimno, Harry Rowe, 21, 30–31, 36–37, 78–81, 107, 160, 188, 192–193
Minnick, Robert, 193
Mitchell, Herbert Francis, Jr., 188
Monroe Calculating Machine Co., 39–42
Monsanto Chemical Co., 233
Multiplication, 87–94
Multiply/divide units, 149

Nagel, Ernest, 170
Nash, John Purcell, 194, 232
Naval Mine Warfare School, 115–120
Neisser, J. B., 115
Nelson, E. C., 166
Newton-Raphson method, 91–93
Nichol, F. W., 82, 109

Observation, knowledge through, 12
Oettinger, Anthony, 208, 213, 216–218, 227, 230, 238–239
Orbits, calculations of, 169–172

Palmer, John, 159, 187, 189–191
Partial products method, 87–88
Patterson, John, 48
Pearson, Karl, 34
Phillips, John George, 82–83
Pierce, George Washington, 21, 110
Planets, orbits of, 169–172
Planimeters, 25–26
Plate accumulators, 96
Porter, William, 162
Price Jones Co., 223–224
"Proposed Automatic Calculating Machine," 53–59
Publication, Aiken and, 217–219
Public utilities, computers and, 213
Pugh, Emerson, 51
Punched-card systems, 64, 149
Punched-tape readers, 151
Purcell, Edward Mills, 23

Pusey, Nathan Marsh, 224, 229
Pym, H. E., 75, 78

Radcliffe College, Aiken and, 25
Randell, Brian, 64, 67
Registers, 97, 147, 149
Relays, 35–36, 40–44, 147, 149, 153–154
Reliability, 40, 153, 156
Retirement from Harvard, Aiken's, 227–230
Robinson, John, 16

Salton, Gerard, 228–229
Salzberg, Bernard, 179
Santamesas, José, 243
Saphire, Larry, 96, 162
Saunders, Frederick, 66
Scheraga, Harold, 208
Scheutz, Georg, 69–70
Security, computers and, 233–236
Seeber, Robert Rex, 162–163, 198
Selective Sequence Electronic Calculator, 122, 163
Sequence mechanisms, 155
Shapley, Harlow, 45–46, 102–108, 119, 125–127, 131, 133–139, 172–173, 225
Sill, Jacqueline, 237–240
Sine-integral function, 160
South Kensington Museum, 68–69
Spengler, Hans, 154
Stanley, John Pearson, 181
Sterne, Theodore, 82, 173
Stewart, Milo, 11
Stibitz, George, 44
Stokes, Rowland, 15–16
Stoll, E. L., 5
Stone, Marshall, 29
Students, Aiken's relations with, 2, 215–217, 238–239
Subroutines, 152
Subtracting, progressive, 90

Switches, 147, 149
Switching theory, 194

Tables
  of Bessel functions, 177–183
  for multiplication, 87–88
  of reciprocals, 93
Teaching, Aiken's, 23–25, 238–239
Telegraph, relays for, 36
Tropp, Henry, 235
Typewriters, 150

UNIVAC computer, 205
University of Chicago, 17–18
University of Wisconsin, 11–14

Vacuum tubes, 27, 40–44, 90
Van der Poel, W. L., 219–220
Verdonk, Frank, 162
von Neumann, John, 164–166

Walker, Arville, 127
Wang, An, 217–218
Watson, Thomas J., Jr., 48, 90, 197–199
Watson, Thomas J., Sr., 47–50, 109–110, 121–122, 125, 128, 131–145
Weaver, Warren, 14, 17, 164, 211
Weinbrenner, George, 241
Westergaard, Harald, 82, 85, 106, 131–135
Westinghouse Electric Manufacturing Co., 13
Whatmough, Josiah, 208
Wheatland, David, 111–112, 160, 180
Whipple, Fred, 172–174
Wilkes, Maurice, 2–3, 5, 175–176, 216–217, 222, 228, 239
Willis, David Grinnell, 194, 232
Woo, Wei Dong, 188
Wright, Thomas, 115–118

Yamauchi, Hiroshi, 244

Zuse, Konrad, 6